Going Back for Our Future II

Carrying Forward the Spirit of Pioneers of Science Education

A volume in
Pioneers of Science Education
Jon E. Pedersen, *Series Editor*

Going Back for Our Future II

Carrying Forward the Spirit of Pioneers of Science Education

edited by

Jon E. Pedersen
University of Nebraska–Lincoln

Kevin D. Finson
Bradley University

Barbara S. Spector
University of South Florida

IAP
INFORMATION AGE PUBLISHING, INC.
Charlotte, NC • www.infoagepub.com

Library of Congress Cataloging-in-Publication Data

A CIP record for this book is available from the Library of Congress
http://www.loc.gov

ISBN: 978-1-68123-191-4 (Paperback)
978-1-68123-192-1 (Hardcover)
978-1-68123-193-8 (ebook)

Printed and bound by CPI Group (UK) Ltd, Croydon, CR0 4YY

CONTENTS

INTRODUCTION

Pioneers: That Pesky Definition

This book is the second in a series about pioneers in science education. In the first volume's introduction, we, and to some extent the chapter authors, rummaged around in our thinking for what we might know about pioneers in science education. We asked who were the pioneers and what made them pioneers. We stated, "Trailblazers and pioneers are individuals who invent new paradigms, are among the first to venture into new territory, to explore new ways of doing things, to rethink old ideas and fashion new ones—to shift paradigms that redefine and redirect life and living" (p vii). That definition seemed to serve our needs adequately at that time. Yet the basic question that has continued to arise in our discussions with colleagues is, "What is the definition of a pioneer in science education?" It seems no matter what audience we've been with or in what format we've presented information about that first volume, we keep being redirected back to that same fundamental question. As well, as we have discussed the definition of "pioneer," questions have arisen about how impact is measured and, to be a pioneer must one have a presence in all dimensions of scholarship (research, teaching, service) or could contributions be in just one area? There have also been questions of what constitutes "science education" as a field. How "deep," if you will, do we probe into the profession itself and consider those who may be more associated with a natural science (i.e., physics, chemistry, biology, earth/space science) and less with education. Given all of these questions, perhaps our first effort was too generic and needed

Going Back for Our Future II, pages vii–x

refining through the lens of science education as a field in its own right. We get the distinct sense that our audiences have not had a clear answer either and are looking for a definition that will satisfy our collective conception of what constitutes a pioneer in our field. We do find it interesting that no one has disputed the idea that pioneers in science education are relevant and worthy of writing about and understanding at a deeper level. In fact, most all audiences and individuals could list many "pioneers" from their own experience or simply from knowledge of an individual's work or presence in the profession.

The struggle with deriving a good-quality definition of "pioneer in science education" is important, if for nothing else than bringing to focus what is truly historically important for our field. Knowledge and understanding about our past can and should inform us about where we are today and help answer why we are there. It can and should inform us about where we seem to be going in the future and help us more clearly map out how to get into that future in order to realize its potential benefits and avoid its pitfalls. We scarcely have the time and resources to continually reinvent the proverbial wheel (but seem to do much of that anyway). So it is really in our collective best interests to know where we come from and what that means for where we want to be in the future. Hence, a book such as this is important and, we hope, very relevant for the preparation of science educators.

When perusing discussions from the field of linguistics, the meaning of "pioneer" emerges in terms of synonyms, hypernyms, and troponyms. A hypernym is a word that has a broader meaning and is superordinate to other words (e.g., "color" being a hypernym to "red"). A troponym is the way the word has a presence or the manner in which it is used in relation to other words in its lexicon (e.g., "nibble" and "binge" are two manners of "eating"). In one sense, "pioneer" means someone who opens up a new area or of preparing the way for others to follow. Its hypernym is to introduce or to innovate. In a second sense, "pioneer" means to take the lead, or to participate in the development of something. Its synonyms are cause, make, do, and initiate. Its troponyms are many: attempt, constitute, establish, found, organize, orchestrate, set about, spark, take off, trigger, and undertake. These terms are ones we have frequently heard expressed in our dialogues with colleagues in science education about pioneers. When we use them in the context of telling the story about a person, these words tell us something about that individual that fits somewhere in our conceptual framework of what a pioneer is. Yet our conception remains incomplete.

Perhaps we need to refocus our question about defining a pioneer from one that is declarative to one that is more operational. What are the characteristics and behaviors one must exhibit to be counted among the pioneers? One can infer some of this from publication records, but that is but only a piece of a snapshot of a person's life and is somewhat limited and potentially skewed in relation to "actual" impact on a profession. In a way, one might view some of those pioneers like a Johnny Appleseed, taking seeds of ideas and planting them in other ground and nurturing what they've planted so that it grew into something fruitful and useful for the rest of us. Their stories are really about what they did in this respect and how they did it (e.g., teaching, working with graduate students, service to the profession, etc.). Those familiar with operational definitions know that they change over time and as contexts change. They change as the learner becomes more knowledgeable about the target and requires increasingly sophisticated descriptors relevant to the definition. Therein lies part of our difficulty with defining a "pioneer." Each of us has learned about key events in our life timelines (or should) with regard to developments and movements in science education. Those events help contextualize how educators of a given time viewed science and education, and contributed to ways ideas were formulated and promulgated that have had long-standing influences on our field. One such example is science inquiry. Since each of us is at a different point in our learning about these things, so too are each of us at different points in what operational definition of pioneers seems useful. So deriving any single definition that works for all of us may not be possible. So we are left with tasking authors with telling us why the person they write about should be considered a pioneer and then support that with the evidence they have gleaned.

In a way, putting together this volume has been more laborious than the first volume. Part of what we've confronted certainly deals with the absence of a crystal clear definition of "pioneer" that authors could then use to construct their chapters and frame the stories they wanted to share. Part of the labor has simply been finding people who are able and willing to write about science educator predecessors. The task becomes more arduous when the writing must zero in on what makes someone a pioneer, when a lengthy listing of publications is not the focus, and when one must do a lot of digging to find out what motivated the pioneer and led him or her to be involved in science education and accomplish the things he/she did. The weight of the task becomes even greater when the first-person accounts are limited or can no longer be accessed. Indeed, what we have found in working on this volume is that an increasing number of the pioneers our authors have identified do not have long scholarly records (although some are breathtakingly significant and comprehensive), but have been deeply engaged in working with people to further an idea or a way of thinking and practice. Consequently, the authors have had an extra homework burden of seeking out those who personally knew the pioneers and could recall personal sharings, conversations, and experiences that could illuminate the core of what we hoped to find.

We are hopeful that what we have been able to include in this volume of the *Pioneers in Science Education* series will be invigorating, enlightening, and help you to arrive at a better understanding of who our science education predecessors really were and how their sacrifices and endeavors have shaped our field. We are optimistic that you will find something that will help you continue your efforts to improve science education. We believe that we have been able to bring more clarity to what our past is and inform us about where we are—and need to be—going.

—The Editors

CHAPTER 1

DOROTHY L. GABEL

A Life of Service

Meredeth McAllister
Butler University

Li-Ling Yang
Roger Williams University

M. Marc Allan
Butler University

A pioneer in chemistry education, Dorothy L. Gabel studied key areas of teaching and learning in science and is best known for her work in the areas of problem solving, the particulate nature of matter, and the development of preservice teachers. Gabel's life was characterized by her faith, enthusiasm, courage, wit, ideals of integrity, service, and loyalty, as well as her devotion to the pursuit of truth.

Going Back for Our Future II, pages 1–39

DOROTHY'S DEVOTION TO SCIENCE EDUCATION AND THE CHURCH

Dorothy Gabel received little science instruction from the teachers at the small Catholic elementary school she attended in St. Louis in the 1940s. Yet she was fascinated by nature and spent her leisure time in activities related to science. At her family's St. Louis home, Gabel collected swimming insects from a backyard pond to create a "zoo with live insects." Neighborhood friends were invited to take part in these adventures. After her brother, Carl, was given a microscope for Christmas, they enjoyed observing items such as plant cells from an onion skin (Cardellini, 2005, p. 9). In those formative years, Dorothy "enjoyed nature as a member of a youth club and gardening as a Ladybird and Girl Scout." She recalled that her "first experience with chemistry consisted in covering copper pennies with mercury, using a chemistry set from a friend" (Cardellini, 2005, p. 9). These nature observations and inquiry experiences as a young child influenced Gabel's adult perspectives about teaching and learning science. In her 1989 article, "Let us go back to nature study!" Gabel discussed the importance of providing hands-on activities for students "to teach children to think" (Gabel, 1989b, p. 728). This was certainly the case for Dorothy herself.

At the beginning of World War II, Dorothy's father, Carl Gabel Sr., became the head of the security force at a munitions plant on Long Island, New York, and later in St. Louis. Both security positions were at factories that manufactured components for military aircraft. Dorothy's brother, Carl Gabel Jr., explained in more detail:

> [He] had a lot of administrative work to do. Plus keeping the spies out. It was wartime, and my father came home every other night. My mother sort of held the fort, and ran the house in his absence and he worked long hours, long days. (personal communication, September 4, 2014)

After WW II, Dorothy's father continued working in the security field at several of the Atomic Energy Commission's various facilities, with his primary office in the Washington, DC, area. The family moved quite often. Dorothy attended three different Catholic high schools in a 4-year period, one of which was Visitation High School, an all-girls school near Chicago, Illinois, which was under the care of the Sinsinawa Dominican Sisters. She was the photographer for the school newspaper and completed courses such as biology, physics, and chemistry. While she enjoyed her physics course, she didn't care for biology, or even chemistry. She received a grade of A for these courses; however, she later stated that she didn't understand the subjects. It wasn't until she was in college and took another chemistry course that she began to enjoy the discipline. "I liked it because there was not much to be learned by heart," she explained. "Instead, there were problems to be solved and in stoichiometric calculations [that I] could [do] pretty well" (Cardellini, 2005, p. 9).

Although the 1950s were a particularly hard time for women in science (Rossiter, 1995), women did make significant contributions to science around that decade. Even earlier, a report published in *Nature* in 1939 detailed Lise Meitner's discovery of the tremendous energy released when splitting a uranium atom (Haber, 1979). Dorothy Hodgkin (Crowfoot) won the Nobel Prize in Chemistry (the fifth woman to win the Nobel Prize in the sciences) in 1964 for her work on the structure of complex organic molecules, including penicillin and vitamin B12, using X-ray diffraction (Abbott, 1984). Rossiter (1995) reported the increasing number of women going into science, citing a fourfold increase of women earning scientific degrees between 1951 and 1957. Even within the realm of Rosary College (River Forest, Illinois), which Gabel attended in the mid-1950s, many of the founding Sinsinawa Dominican Sisters who helped move the college from Wisconsin to Illinois in the early 1920s had earned doctoral degrees in their fields of expertise. A fair number of these degrees were in the sciences. Around the time that Gabel was a student, there were 12 faculty members in the Natural Sciences and Mathematics Department. Of these faculty members, 10 were sisters, and of these, 4 had already earned PhDs, 2 were doctoral candidates, and the remaining 4 had pursued graduate study after earning their master's degrees. Of the 12 faculty, 3 taught biology, 3 chemistry, 2 geology and geography, 1 physics, and 3 mathematics.

As explained by Rosary College archivist Steven Szegedi, "While it may have been unusual for women to pursue science degrees, much less be encouraged to do so, you will find that women who attended Rosary College would find such a suggestion much more common" (personal communication, September 18, 2014). The increase in "scientific womanpower" was expected to be channeled into acceptable teaching positions in elementary and secondary school science (Rossiter, 1995), and this was the case for Dorothy. She graduated in 1957 with an AB in chemistry (Dominican University, 2014) but decided that teaching was to be part of her life's work. After college, Gabel also made the decision to devote her life to the church. She became "Sister Dorothy" within the Sinsinawa Dominican Sisters of the Most Holy Rosary religious community. Rosary College "had a responsibility to provide teachers for more than 100 elementary schools, 14 high schools and two universities in the United States" (Cardellini, 2005, p. 9). Thus, even though Gabel's degree was in chemistry, she took courses of instruction in teaching and had some practice teaching before she served as a full-time chemistry teacher at Edgewood High School, a Catholic school in Madison, Wisconsin, from 1959 to 1966. Located not far from the University of Wisconsin, Edgewood High School was one of three schools, elementary, middle, and secondary, located on the campus; of the student population, 800 were high school students. There were 90 sisters who lived on the campus and innumerable staff members who supported what was then the only Catholic school in the area. Sister Joan Brichacek taught high school science with Gabel at Edgewood in the late 1960s. She explained that each day was full of activity, starting very early with prayers and breakfast, and the school day from 8:15 to 3:00, followed by after-school activities with students. According to Brichacek, "In the high school, Dorothy was right next door to me, so we shared a lot. But there was also a lot of collaboration among the three buildings there" (personal communication, August 24, 2014). The principal, Sister Kathleen O'Connell, was supportive of her teachers, and according to Brichacek, the sisters had a lot of curricular freedom: "She let us experiment with anything new and innovative that we wanted to try" (personal communication, August 24, 2014).

Gabel and Brichacek were young—barely 4 years older than the students they were teaching—and conditions were far from ideal. "No air conditioning in those days," Brichacek said,

> We would stand next to the open windows [in a full habit] when we taught in the afternoons. And you have to enjoy what you're doing, which we both really did. We were in an environment that was conducive not only to learning but being good to one another. You know [the faculty] had lots of get-togethers, sometimes at the family farmstead of our sister principal [laughing]. Everyone would gather together and share. (personal communication, August 24, 2014)

Edgewood High had a rigorous science program. Students were well prepared to enter college upon graduation. Gabel taught chemistry as well as a general science and was able to give a lecture one day a week, with an average of 40–45 students in the class. This was followed by small groups that met during a 40-minute period. Laboratory experiences were extended to a 60- to 80-minute block and were limited to twenty students per lab to ensure a quality learning experience with available materials. As Brichacek explained, "The hands-on part was the most interesting." "Dorothy had many students in chemistry that went on to join the medical profession or research in the field of chemistry," Brichacek said. "Many students got a good background in science and so they pursued a career in one of those fields, which was so rewarding for the teachers" (personal communication, August 24, 2014).

George DeBoer, a noted scholar of scientific literacy, believed "it was the perceived poverty of scientific knowledge itself that motivated most science educators in the decade of the 60s" (DeBoer, 2000, p. 587). He described,

> [the] approach to science teaching that was emerging was characterized by a more demanding treatment of the academic disciplines, and was based on the belief that within the science disciplines lay the knowledge that would put our country on track both economically and militarily. The new science courses, designed in large part by scientists themselves, were academically rigorous, and special efforts were made to attract exceptionally bright students to study science. (p. 588)

These new courses represented an "alphabet soup" of programs, including *Science—A Process Approach* (SAPA), the *Science Curriculum Improvement Study* (SCIS), and *Elementary Science Study* (ESS). Each program had its own scope, sequence of ideas, and attention to the development of science skills. The instructional approach for programs such as SCIS varied. Through a study of historical records, DeBoer (2000) determined that

> the preferred pedagogy was an inquiry approach, not to develop independence of thought as 19th century scientists had argued, but to mirror and thereby appreciate the way scientists themselves did their work. The main goals of the science community were the preparation of future scientists and a general public that would be knowledgeable enough to be sympathetic to the work of scientists. (p. 588)

While Gabel was teaching high school between 1964 and 1969, she attended Purdue University pursuing a master's degree program and a National Science Foundation Summer Institute in chemistry. The decline in math and science college graduates during the 1950s prompted the NSF to take action, and in 1954 it began to fund summer institutes for K–12 teachers. The program began with one summer institute that year, but the NSF precollege teacher education programs grew rapidly and quickly, with the budget increasing from $150,000 in 1955 to a high of $125 million in 1966, 1967, and 1968. At the start, the institutes were intended to increase the subject-matter knowledge of individual teachers or to offer experienced teachers advanced graduate work (Lagemann, 2000, pp. 168–169). Many science teachers took advantage of the NSF program. Fellow science educators John Staver (professor of science education at Purdue University) and Bob Sherwood (professor of science education at Indiana University) both met Gabel at Purdue where they were participating in one of the NSF institutes. According to Staver, "Teachers had the option of focusing on one of several disciplines available—chemistry, biology, physics, and I think mathematics. And you could do this program for four summers in a row, or a summer and an academic year, and then the next summer" (personal communication, August 13, 2014). Gabel received a master's degree in chemistry from Purdue in 1969. Jane Kahle, a renowned science educator who was a doctoral student at Purdue in the early 1970s, described Gabel as a very good chemist as well as a science educator. Kahle explained, "The truth is that every good science teacher from that era if they could get into these teaching institutes, very few went back. We all stayed and got done with our PhDs" (personal communication, September 2, 2014). She elaborated further on the significant role that the NSF institutes played in preparing science educators with robust content background, stating,

> Those teacher institutes were really critical because we did them [and] the idea was to give us current science content knowledge. I mean, we took graduate science, chemistry, biology, and physics. So that idea [the importance of a solid content background] was reinforced because of those experiences. And those were pivotal experiences. That particular program paid off for this country hugely in educating a whole generation of science educators. (personal communication, September 2, 2014)

Replicating the experience of many science educators, Gabel returned to Purdue in 1971 as a full-time graduate student to obtain her doctoral degree. "Although I liked to teach chemistry in high school," she said in a 2001 interview (Cardellini, 2001), "I thought I could give the community a more important contribution working on the training of future teachers of chemistry, which is not taught at the high school level" (Cardellini, 2005, p. 9). Gabel joined the graduate program in biological sciences at Purdue. As Kahle explained,

> One of the interesting things was that we all had joint appointments if we entered the faculty, or like Dorothy and I, when we were graduate instructors, because the science education was always placed within the disciplinary departments. So my appointment was in biological sciences and hers was as well. It was very unique. (personal communication, September 2, 2014)

The Biology Department had grant funding to provide Gabel the support she needed to begin her graduate studies.

Soon Gabel met an educator who was to become an influential person in her life and career, J. Dudley Herron. In a 2001 interview, Gabel described Herron as "the first big influence" in her life as a teacher in the sciences. "It was as if we share ideas constantly, and I knew I had a guide who supported me as a person and as a colleague" (Cardellini, 2005, p. 9). In the same interview, Gabel said that Herron suggested she consider the possibility of doing a PhD in science education. "I was very impressed with his enthusiasm for teaching, his deep religious convictions, and his loyalty towards me" (Cardellini, 2005 p. 9). Purdue continued to offer support for Gabel's PhD studies during her second year—an instructor position with the Chemistry Department. She assisted Herron in teaching the chemistry teaching methods course and supervising student teachers. That year, she also became

Associate Director of the Chemistry Institute. Gabel stated in a 2001 interview, "It was during this year that I really knew Dudley closely and I realized I had made the right choice when I decided to study with him" (Cardellini, 2005, p. 9). Gabel and Herron shared similar teaching philosophies, with the overarching goal to make the teaching of chemistry in high school conceptually understandable for students. As Herron described what they shared in common, "Dorothy and I are alike in many ways. We are about the same age, we have similar values, and our intellectual curiosity has been directed toward religious faith as well as chemistry and teaching" (Herron, 2009b). During Gabel's third year at Purdue, Herron went to Malaysia for the entire year to act as a regional director for a science center. Gabel assumed his teaching duties at Purdue and designed the facilities for chemistry education in the new chemistry building while also mentoring graduate students, such as John Staver. When Herron returned during Gabel's fourth year, she supervised classes for the Department of Chemistry, including the course Chem 100, an introductory course taken primarily by engineering students. At the end of her doctoral program, however, Gabel had some difficult decisions to make. She and Herron talked a great deal about whether she should accept the position she had been offered at Indiana University or accept a position at a college supported by her order, Edgewood College in Madison, Wisconsin (where she later served as a trustee between 2000 and 2006); or teach instead at Dominican University (the renamed Rosary College) in River Forest, Illinois. Herron believed that Gabel was "uniquely qualified with the research expertise and the understanding of the issues to go to a secular institution and make a contribution" (personal communication, August 13, 2014). In a letter to one of Gabel's Dominican Sisters, Sister Mary Joan Gibson, Herron expanded:

> She knew that she would miss the companionship and support of other sisters, but the central question was where she could serve God best. At the time we both saw her outstanding potential for research in chemistry education, and the question of where she could best apply those talents was what tilted her toward work in a secular institution rather than a religious one. (Herron, 2009a)

Sister Angelo Collins, an accomplished science educator and Gabel's fellow sister in the Dominican order, explained,

> Before 1974, if you were looking at the Sisters of Sinsinawa Dominican, 90 percent of us were teaching in Catholic schools. By 1978, we were doing a lot of different things. 1974 was in the middle of a period of great unrest among nuns. The Sinsinawa Dominicans were in the process of reform, so the Catholic Church said to us go back to your "founding" and figure out what you were founded for, and then redirect your efforts in that way. So it was a period of great tumult. (personal communication, August 24, 2014)

As Herron explained,

> [should] she accept the position IU had offered, where she would certainly teach, but where research was expected as well. In what capacity could she be of greater service was the central issue, but there were personal considerations as well: At Edgewood or Dominican she would be with family—her Dominican Sisters; at IU she would live alone on an allowance provided by her Sisters while her comparatively robust salary would go to support the family as a whole and its broader mission. If Dorothy ever regretted her decision, it never showed. (Herron, 2009a)

Whenever possible—during summers, while attending professional meetings in Indiana, New York, New Jersey, Illinois, Wisconsin "or any other spot inhabited by Dominican Sisters, and when the sisters could use her professional services—Dorothy visited with and worked for 'her family'" (Herron, 2009a).

THE PROBLEM BEHIND THE PROBLEM

Herron's influence on Gabel's work was in her application of Piagetian thought and developmental levels to her research in problem solving. While most studies were concerned with a "mismatch" between developmental level and the complexity of chemistry problems, Gabel believed that many chemistry topics, such as stoichiometry and molarity, required formal-operational thought (Gabel & Bunce, 1994, p. 314). Herron described the circumstances under which their thoughts around problem solving evolved as following:

> A lot of what we did with problem solving were [from] seminars with math education. There was more collaboration here between math and chemistry than any other groups. The mathematicians were a little further ahead of us, as the physicists were further ahead of us when it comes to cognitive deficiencies or gaps in their knowledge. But we were looking at those kinds of things and that kind of naturally evolved into the problem solving, and all of that grew out of Piaget's work. She was always saying look for the "problem behind the problem." She was going to a deeper level than a lot of research that was going on at the time she was practicing. And certainly her work was saying you have to understand how people learn. (personal communication, August 13, 2014)

Between 1972 and 1976, research articles in volumes 56–60 of *Science Education* illustrated the trend toward the application of psychological theories of development and learning from Piaget to problem solving in science education. These articles helped define the component skills of scientific problem solving as the "mental operations overtly manifested in the ability to conserve or exclude irrelevant variables." In the theoretical terms of instructional psychology, these component skills—observation, description, and measurement—are called "process skills" (Champagne & Klopfer, 1977, p. 446). This body of work was a foundation for the reconsideration of problem solving as an instructional method and, at the same time, "the better definition of problem solving skills makes it possible to devise more precise and more focused procedures for assessing problem solving behavior as an outcome of science instruction" (Champagne & Klopfer, 1977, p. 446).

Gabel completed her doctoral thesis, an examination of an entire year on the effect of study according to learners' own pace compared to study with deadlines, and work with others compared to working alone, on learning rate, retention, and attitude in the sciences. She utilized a sample of 1,022 seventh-grade students enrolled in the Intermediate Science Curriculum Study (ISCS) program (Cardellini, 2005). Gabel knew this program well, having worked with Herron on ISCS activities at Purdue (Howe & Early, 1979). Gabel published her dissertation study, "The Effects of Grouping and Pacing on Learning Rate, Attitude, and Retention in ISCS Classrooms," in the *Journal of Research in Science Teaching* (JRST), with Herron as co-author (Gabel & Herron, 1977). The findings shed light on rate-of-learning studies, an area for which research had not been conclusive. This paper was highly acclaimed and received in 1978 the JRST Award, which is given annually by the National Association for Research in Science Teaching to the article that is judged the most significant publication for that year.

The Piaget developmental framework was evident within another Gabel paper, "The Development and Construct Validation of a Group-Administered Test of Formal Thought" (Staver & Gabel, 1979), which she co-authored with John Staver, her first graduate student at Indiana University. Staver called Gabel a major influence on his development as a graduate student. He described their relationship as "a very traditional, classical relationship between teacher and student" (personal communication, August 13, 2014).

"Obviously [I did] a lot of reading about Piaget and what the formal operations were about and what tasks to use. And I would read about things and share them with Dorothy," Staver recalled. He further stated, "She was always very supportive, and meticulous. I mean really meticulous [laughing]. She had a great grasp of details and shared those with students just as much as the big picture" (personal communication, August 13, 2014). At the time, Staver said, there was a lot of drive for paper and pencil instruments (personal communication, August 13, 2014). The thesis, published as an article in JRST in 1979 with Staver as first author, received the JRST award in 1980 for its significant contribution to developing "a reliable, valid, efficient, and practical measurement tool," the Piagetian Logical Operations Test (PLOT), a prerequisite to the process of matching "instruction and curriculum materials with the developmental level of the learner" (Staver & Gabel, 1979, pp. 543–544). As great a researcher as Dorothy was, Staver said, she did not hold it separate from teaching and service. "They are very much connected and combined and maybe even unified" (personal communication, August 13, 2014).

Gabel believed that research should inform teaching practice. Raising the level of student achievement through chemistry education research was a major goal. She became active in the Hoosier Association of Science Teachers, Inc. (HASTI), and between 1978 and 1980 she served as president. In her HASTI presidential address, she discussed her beliefs about problem solving. With the rapid increase in knowledge and the limited time that teachers have to introduce and discuss concepts, she thought teachers and students therefore more often rely on a memory approach. Beginning with a \$5,000 grant from the Spencer Foundation and a larger NSF grant, \$52,000, awarded in 1979, Gabel studied high school chemistry students' memory structures and how they could be used in rule applications relevant to problem solving (Gabel & Sherwood, 1983). Certainly an increased emphasis on statewide testing encouraged memorization over understanding. She argued that while some U.S. states and school districts interpreted the national standards for the teaching of science (National Science Education Standards, NSES) as a minimum requirement, many scientists felt that this minimum was insufficient, so they added additional standards. As a result, "many students resorted to memorization rather than understanding" (Cardellini, 2005, p. 10). Gabel explained in a 1981 press release, "Too much emphasis on memorizing of facts rather than learning concepts is making robots of many school children" (Indiana University, 1981). Gabel maintained an active presence in the organization throughout her career and was awarded the Distinguished Service Award from HASTI in 1996.

Gabel's work using algorithms in problem solving is well established. She explained, "Evidence indicates that many students' conceptual understanding is inadequate to solve chemistry problems that require transfer and that students frequently solve 'exercises' using an algorithmic approach almost exclusively" (Gabel & Bunce, 1994, p. 304). In a study conducted in 1981, Gabel set up a learning approach using common strategies, including the factor-label method, diagrams, or analogies. She determined not only that the majority of students relied on strictly algorithmic techniques, but that their lack of conceptual understanding, and hence lack of use of concepts in solving problems, were shown by the fact that many more students could solve problems successfully than could answer the interview questions about the concepts involved (Gabel & Bunce, 1994). Gabel continued this work with Sherwood. Their collaborative research on the relative effectiveness of four instructional strategies (factor-label method, proportionalities, analogies, and the use of diagrams) on the problem-solving performance of high-school students was, again, directly related to improving the level of student achievement in chemistry. The factor-label method was observed as the best instructional choice of the four options, and the proportionality method the worst (Bodner & Herron, 2002). This research study, "Facilitating Problem Solving in High School Chemistry," published in JRST in 1983, won the National Association for Research in Science Teaching (NARST) Research Application Award (Gabel & Sherwood, 1983). This was the second NARST research application award that Gabel received, from a total of five awarded for her work in problem solving from 1981 to 1987.

In 1984, Gabel, Sherwood, and graduate student Larry Enochs continued the study of high school students' problem-solving skills, noting that students with a strong proportional reasoning ability were more likely to use algorithmic reasoning than those students with low proportional reasoning ability (Bodner & Herron, 2002). This collaboration was a fruitful one, setting the foundation for their work and forming the basis for problem-based instruction. Sherwood stated, "She worked so hard and was so concentrated, and I think she saw the idea of creating new knowledge was going to help people" (personal communication, August 14, 2014). Sherwood's office was closer to Gabel's than Enochs's was. Occasionally, he said, he would hear Gabel yell for him. "I would go over and she would say 'Let's redo this and this and this.'" She did value the input from the graduate students, but she had definite ideas about how things should go and was pretty straightforward about that." Gabel, he said, taught him the importance of good research design and well-designed instruments, because such instruments will tell researchers what methods to study (personal communication, August 14, 2014).

Gabel believed that the basis of problem solving is first achieved by understanding the concepts within the problem. When instruction adds more concepts to the problems (such as acids, bases, solubility, and equilibrium), the problem becomes more complex and students apply algorithmic methods to solve them. "The use of algorithms to solve problems without understanding results in students' inability to solve the more complex problems and is probably responsible for student perception that success in chemistry is not necessarily within their control" (Gabel & Bunce, 1994, p. 314). Her work with high school students also included the use of analogies as an instructional strategy to assist in concept development and problem solving. Gabel's research in this area focused on the use of word analogies for fundamental or familiar tasks within chemistry (Gabel & Samuel, 1986; Gabel & Sherwood, 1984). In her 1986 study with co-author Samuel, she argued that a major reason why students have difficulty solving chemistry problems is because they don't understand the underlying chemical concepts. Using a dilution of orange juice as an analog for chemical dilutions, Gabel and Samuel observed that the analog was helpful when the problems presented to students did not involve changing concentrations. However, when confronted with more difficult concentration and dilution problems, students' performance did not improve (Bodner & Herron, 2002). She concluded, "Student performance on certain types of problems may improve as the connection between the analogue and problem is better understood" (Gabel & Bunce, 1994, p. 318).

The Gabel and Samuel 1986 study on solving molarity problems furthered the discussion about connections between students' understanding of concepts and their abilities to solve problems. Using chemistry molarity problems and equivalent analog problems involving dissolving lemonade, they tested 400 high school chemistry students using a multiple-choice format before and after instruction. Results indicated that for simple problems, students were able to solve both the analog and molarity problems. However, when the problems became more complex, such as when the solvent was evaporated or additional solvent was added, fewer students (and in about the same proportion) could solve the analog and chemistry problems correctly. Even after chemistry instruction, there was little improvement in students' ability to solve the more complex analog problems, indicating that the students' difficulties were related more to not understanding solutions than to a lack of understanding of the mole concept (Gabel & Bunce, 1994, p. 312).

In her 2001 interview with Cardellini, Gabel provided an example of the challenges learners have in solving problems. She told a story about asking students to consider how many cookies they could make when they had varying amounts of ingredients such as eggs and oil. About 92% of the students came to the correct solution. She then asked them to solve problems involving moles and masses of less familiar materials, which are generally called chemical reagents. The percentage of students who came to the correct solution was approximately 32%. "The major differences in the problems are not the mathematical difficulties, because they both required the same skills," (Cardellini, 2005, p. 10) Gabel concluded. "It was the lack of familiarity with the substances and the units of measurement used" within the problem. "It seems to me that this happens because we use too many chemical materials and units of measurement within unfamiliar problems in chemistry," she said.

> For some students, even mass and volume are new concepts. And the problems contain unfamiliar units such as grams, liters, and meters. What emerges is that students are frustrated by so many new terms, which leads them to treat the problem [differently] and get an answer without thinking about what they are doing. (Cardellini, 2005, p. 10)

Gabel's work in problem solving led to the identification and corresponding need to address common student misconceptions during instruction.

She believed that emphasizing a qualitative understanding of the basic concepts such as density and volume, followed by the clarification of misconceptions, was important for successful problem solving in the classroom. For example, many students can calculate the density of a solid, yet when shown samples of identical mass but different volumes, they are unable to serially order the samples by density. It is unlikely that having students solve numerous density problems by substituting values into the density formula will help them distinguish between density and volume (Gabel, 1995a, p. 212). Gabel thought students should be able to think critically about a problem, and that meant being able to analyze problems carefully, explain each step, determine what information is superfluous, and add any additional information needed to solve the problem. But the basis of problem solving has always been the ability to understand the concepts. And Gabel's research indicated that students were not grasping concepts.

This is an important connection between Gabel's areas of research: the particulate nature of matter and problem solving. Gabel investigated the particulate nature of matter as a natural progression of her research in problem solving. Knowledge of the particulate nature of matter is also important in solving chemistry problems, Gabel said. Gas law or stoichiometry problems can sometimes be solved without considering the particulate nature of matter, though other chemistry problems cannot. Yet voluminous evidence suggests that many chemistry students lack this conceptual framework that is so fundamental in the study of chemistry (Gabel & Bunce, 1994, p. 308). "Dorothy's extensive research on problem solving, misconceptions, and the importance of macroscopic/particle/symbolic representations," Stacey Bretz, a well-known chemistry educator, said, "are internationally regarded as seminal works in the field of chemistry education research" (Bretz, 2006).

THE PARTICULATE NATURE OF MATTER: PICTURES IN THE MIND

While working on the "problem behind the problem," other research questions emerged concerning the particulate nature of matter. Gabel was responsible for bringing the concept of Johnstone's triangle of symbolic, macroscopic, and particulate representations (Gilbert & Treagust, 2009) to the attention of the chemistry education community. Her interest in the particulate nature of matter and the corresponding visualization of the macroscopic and symbolic representations developed over three decades of study as a natural progression of her work in problem solving. Gabel explained, "Although some stoichiometry problems can be solved without considering the particle makeup of matter, molarity problems and gas-law problems can be understood more readily by making use of the kinetic molecular theory" (Gabel & Bunce, 1994, p. 308). Gabel coined the phrase "pictures in the mind," and it is now common practice to use drawings of particles in textbooks for high school and college chemistry (Gabel, 1983). Gabel's research helped to identify the discrepancy that while experts describe most phenomena in terms of particle behavior, novices give only macroscopic descriptions. She explored various strategies to help resolve the discrepancy and inspired others to search for ways to increase student understanding of the particulate nature of matter (Gabel, 1993b).

In the book chapter "Prior Knowledge of Chemistry Students," Gabel and her co-author, Karen Stucky, who was an elementary school science educator, explained how the various levels may inform teachers of grades K–9 (Gabel & Stucky, 2008). They said that although the levels are displayed on an equilateral triangle, teachers should not infer that the three levels are equally easy to understand nor should they be taught to students of all grade levels. "Studying chemistry at the macroscopic level," they wrote, "is the easiest level to comprehend because this level is less abstract and more accessible, depending extensively on the use of the senses" (Gabel & Stucky, 2008). Gabel and Stucky (2008) concluded that the other two levels of Johnstone's triangle (particulate and symbolic) are considerably more abstract and are not developmentally appropriate for most children in grades K–4 (and even in grades 5 and 6, and for some children in grades 7–9). When topics such as atoms are introduced in grade 3 (as they are in one textbook series approved for adoption in some states), "many children will memorize definitions. Even some teachers are challenged to explain the difference between an atom and a molecule. The complexity of understanding chemistry at the particulate level is simply inappropriate for elementary children" (p. 131).

Conceptual errors can occur when studying abstract concepts. Abstract concepts, such as atomic theory, are often presented to students in the lower grades. However, Gabel reminded teachers (Gabel et al., 1987) that they must live with the reality that "the microscopic level is depicted at the elementary level," so it's not as if teachers can avoid the atomic theory. Gabel, and others, have observed that elementary children are in what Piaget calls the concrete operational stage of development. The implication of this for teaching chemistry at this level is that "children need to work with concrete objects rather than learn about models and theories" (Gabel, 1998). Passionate about making a change in chemistry instruction, Gabel began writing and contributing to the development of textbooks that incorporated instructional strategies based on educational research. Gabel (1999) explained,

> The primary barrier to understanding chemistry, however, is not the existence of the three levels of representing matter. It is that chemistry instruc-

> tion occurs predominantly on the most abstract level, the symbolic level. A cursory examination of almost any introductory chemistry textbook supports this assertion. (p. 549)

She wanted to improve the scope and sequence of chemistry concepts, address the use of specific vocabulary, and reduce the number of new concepts found within textbooks (Cardellini, 2005, p. 10). So she began writing *Chemistry: The Study of Matter* (Dorin, Demmin, & Gabel, 1989) between 1987 and 1989 while working at NSF as the director of the Division of Teacher Preparation. Her brother, Carl, explained,

> Yes, she tells the story that she wrote her chemistry textbook as she was traveling back and forth on the subway and bus when she was at NSF. And I don't doubt it, because she was very studious and could concentrate on the bus. (personal communication, August 2014)

An additional goal in writing chemistry texts was to decrease the presence of misconceptions. Conceptual errors are commonly found in the texts that students use. And what is presented in instruction may be beyond their developmental level. Similarly, misconceptions may arise from words that are used informally in everyday usage but have a more strict meaning within science. Gabel stated, "When science books are not carefully written they may contain statements such as 'heat moves up the rod' or 'molecules expand on heating'" (Gabel, 1989a). Gabel believed that textbooks should utilize the macroscopic view of matter before introducing changes at the microscopic level (and only then representing matter symbolically). She explained, "It would also mean having to find the solution of problems using algorithmic methods (such as the method of factor-label) in a more conceptual (multiplication and division) manner" (Cardellini, 2005, p. 11). She also believed that there is a need to move the concepts focused on the atomic structure of matter to the end of the text.

"The difficulty in writing a textbook lies in the fact that a person with a deep understanding of the structure of matter writes the text and so this structure informs the order of the content," she said (Cardellini, 2005, p. 11).

> Instead, the author of the text should get into the mind of the student, a novice that does not know the structure of matter, and imagine what would be the best way to facilitate the acquisition [of the concepts]. This means starting with something that is familiar to the student and then [address] the abstract structure [of matter so that it is seen] as a clarification of familiar phenomena. (Cardellini, 2005, p. 11)

There are other challenges to writing textbooks—some out of the hands of the author. Textbooks are certainly a negotiation between author, public, and press, among others. Gabel expressed frustration with the book industry's approach to textbook marketing and the development of content based upon outdated recommendations from high school teachers rather than educational research. "Textbooks are a commodity from which the book industry makes a profit," she said. "Probably 95 percent of the publishing houses are not willing to take a risk in publishing a book based on the recommendations of the research" (Cardellini, 2005, p. 11).

Regardless, Gabel was able to incorporate educational research into her writing of chemistry textbooks. In her 2001 interview she stated that the results of the research in which she had been involved did make a difference in the content of textbooks. "But only because I have personally worked with an author, or I was the author" (Cardellini, 2001). The first textbook improvement, she said, was the increased use of the particulate nature of matter reported across the length of the text, rather than being confined to the chapter on molecular kinetic theory. Gabel said she started a study with Cliff Schrader, a fellow textbook author, on students' understanding of the particulate nature of matter using particle drawings to represent all kinds of chemistry changes. As a result of these studies, when Schrader was asked to be an author of the Heath text *Chemistry* with Dudley Herron and others, they included drawings of particles throughout the text. Similarly, when Gabel became a co-author of the Prentice Hall text *Chemistry: The Study of Matter* (1989), she included particle drawings in all of the chapters. Diane Bunce, professor of chemistry at Catholic University of America (and co-author of the book chapter "Research on Problem Solving: Chemistry" [1994] in Macmillan's *Handbook of Research on Science Teaching and Learning* [Gabel, D. L., & National Science Teachers Association, 1994]), collaborated with Gabel on the initial research that led to using particle drawings in the textbook *Chemistry: The Study of Matter* in 1989. This work, using particulate-level explanations in the teaching of chemistry, was continued with the development of the Symbolic-Application-Particulate (SAP) Test to evaluate students' understanding of matter on all three levels (CHED, 2008).

Gabel emphasized her commitment to bridging chemistry research and chemistry education in a 1999 article, "Improving Teaching and Learning Through Chemistry Education Research," in which she advocated for incorporating current research findings into the teaching and learning of chemistry (p. 553). Her research areas of problem solving and the particulate nature of matter have multiple applications to the teaching of college-level introductory chemistry courses. Gabel and Samuel concluded their 1987 study by stating that an increased emphasis on the particulate nature of matter in introductory chemistry courses and the careful representation of particles by chemists when they are used in instruction might bring about not only an increased ability to solve chemistry problems, "but it may also help make chemistry more understandable by providing the framework underlying the discipline" (Gabel & Samuel, 1987, p. 697). In addition, Gabel proposed that instructors help students learn the familiar macroscopic world first, "and then explain that world using the particulate nature of matter." In this way, new concepts are anchored on the concepts of a preexisting network in long-term memory (Gabel, 1999, p. 551). Another approach suggested by Gabel is to utilize instructional strategies that promote the social construction of knowledge. She believed that students construct new understanding only after they have considered their current understanding. This is facilitated through social interaction with their instructors, other students, or perhaps a computer. "This interaction can be concerned with the introduction of conflict situations that help students create dissatisfaction with their current views, modification of current views (partial restructuring) using concept substitution or analogies, or reflection on the meaning of the text through self-explanation" (Gabel, 1999, p. 552).

Throughout her career, Gabel clearly identified common barriers to learning chemistry. Among them, the abstract nature of chemistry concepts; the presence of misconceptions and the difficulty of conceptual change; the complex nature of the discipline with its threefold representation of matter; students working with unfamiliar vocabulary; materials; the improper use of language, such as using familiar expressions that have different meanings when applied to chemistry; the reliance on algorithms to solve chemistry problems; and even the structure of the discipline, such as the sequence of concepts from abstract atomic theory to the macroscopic level (Gabel, 1999). Gabel was awarded the NARST Distinguished Contributions to Science Education through Research Award in 2008. Her significant research into how analogies, models, and visualization enhance learning in chemistry provided a theoretical base for the research of her colleagues. Gabel's research also addressed critical issues in teaching and learning science for elementary and middle school students. She applied her research to solve practical problems in teaching introductory chemistry to nonscience majors, and, as explicated on the award plaque, "has unselfishly and consistently used her vast knowledge of science education research to improve science education for underrepresented students, particularly those in inner city schools."

What sets her research apart from other prolific scholars in chemistry education research, Bretz said, was "her determination to not just investigate questions that are intellectually interesting, but to transform the results of this research into curricula and classroom practice" (Bretz, 2006).

TEACHER EDUCATION

Educating teachers, Gabel discovered, offered an opportunity to apply research to practice. She focused her efforts on two key areas within preprofessional development—the development of content knowledge and pedagogical content knowledge (PCK). She believed that teachers should have a strong science content background, as well as sound pedagogical tools, to more effectively teach science. Gabel's previous studies with preservice elementary education students revealed gaps in content knowledge. For instance, in a 1984 study with Larry Enochs (published in an article titled "Preservice Elementary Teachers' Conception of Volume"), Gabel studied students' conceptions of volume and surface area. A test, using drawings of several objects, asked students to select from a list of possible methods by which the volume and surface area could be determined. Even after instruction, Gabel observed that a large percentage of students did not understand either concept nor could they distinguish between the two. While most students recognized the volume of a rectangular solid, as a product of length, width, and height, only 44% could determine the volume as a product of area and height. Again, Gabel observed that most students will solve problems using a "'memorizing mode' rather than a 'thinking mode' to solve the problems" (Gabel & Bunce, 1994, p. 310). Gabel and Enochs encouraged other science and science methods instructors to "be aware that many preservice elementary teachers do not understand the concept of volume, and therefore they must teach the concept instead of assuming that it is already understood and self-evident to all students" (Gabel & Enochs, 1984, p. 679).

In Gabel's 1987 study on preservice elementary teachers' conceptual understanding of the particulate nature of matter, she determined that although there were significant differences on some aspects between those preservice teachers who had taken a chemistry course versus those who had not, both groups had an overall inadequate understanding (Gabel & Bunce, 1994, p. 308). And even though Gabel believed that the microscopic level should not be taught in the lower grades, it was nevertheless present in the elementary curriculum and thus could not be ignored in an elementary methods course.

> It is the reality that the microscopic level is depicted in elementary science texts, she said. This, coupled with the fact that elementary teachers would be familiar with the particulate nature of matter to make sense of everyday phenomena that they encounter and teach to children, explains the inclusion of the topic as a part of a Basic Science Skills course at Indiana University. An additional reason for inclusion of the topic was to show students how theories and models are an outgrowth of the other science process skills of observation, inferring, predicting, hypothesizing, experimenting, etc. (Gabel & Samuel, 1987, p. 695)

In addition to the development of content knowledge and the practice of using process skills as a part of PCK, Gabel sought to show how chemistry was relevant to students' lives. She stressed the need for teachers to be able to establish links between content and daily life while actively involving students in the conceptual understanding of the content. Gabel had observed that elementary preservice teachers had "little conceptual understanding of basic phenomena such as burning, melting, decomposition and dissolution" (Cardellini, 2005, p. 11). When teachers understand the concepts and the developmental needs of learners, she said, these common ideas can motivate students. However, it is often the case that preservice teachers do not understand these common ideas. They might come to an undergraduate science methods class with bits and pieces of chemistry knowledge or no chemistry background at all. Gabel's concern was that "if chemistry is presented by these teachers to children, it is frequently taught in an unorganized manner," as a demonstration, or simply read to the elementary students (Gabel, 1985, p. 703).

In the summers of 1984 and 1985, Gabel was able to pilot a new sequence of teaching chemistry concepts and skill development through two week-long summer programs for gifted children at her alma mater, Rosary College, in River Forest, Illinois (Gabel, 1985). These experiences gave Gabel an opportunity to fine-tune the pedagogical approaches she wished to use in her own course (Q200, Introduction to Scientific Inquiry) for elementary-education majors. The approaches incorporated her beliefs that chemistry is best learned by seeing and doing rather than by reading and listening, and that inquiry strategies should be used to teach concepts and motivate children to study chemistry on their own rather than convey the idea that "chemistry is magic." She stressed the importance in instruction of using both simple and integrated science process skills. Chemistry concepts build on one another, she said, and the macroscopic properties of matter should be viewed as more important than microscopic properties for children at the elementary level (Gabel, 1985).

During those two years, the Q200 course transitioned from an instructional model of lecture and lab to a lab-only format. Colleagues and graduate students in the college encouraged Gabel to make this transition. This also became an opportunity for Gabel to mentor graduate students in a new way. Former graduate student and professor of science education at the University of Nebraska Larry Scharman explained,

> It allowed the graduate instructors, these people that are certified and experienced teachers, to have an opportunity to develop that relationship with the students. It also allowed me to converse with Dr. Gabel and reflect on how things were going in the new format. (personal communication, August 15, 2014)

This course reform was an acknowledgement of the graduate instructors' expertise and emerging knowledge of the connections between research and practice. Scharman described this as "a breakthrough." "It was allowing me certainly not to be a peer, but at least break down a barrier and make meaningful contributions to what she was doing. So there was much more respect and trust" (personal communication, August 15, 2014).

Over the years, Gabel offered many graduate students the opportunity to teach the Q200 course. She worked closely with the graduate instructors to ensure that everyone was well prepared to teach. She provided directions or suggestions on how to prepare each lesson during biweekly meetings, Ling L. Liang, a former doctoral student of Gabel's, said, and she also typically prepared the demonstrations for her own lectures a week before, so she would always be prepared for everything in advance. "Prior to working with Dr. Gabel, I had been teaching traditional lecture courses, demonstrations, and confirmation labs. My initial understanding of inquiry really came from Gabel and other science education faculty members at IU" (personal communication, August 12, 2014). Teaching Q200 allowed the instructors to reflect on their own learning as well as their students' learning. "I did more learning about student knowledge characteristics, misconceptions, and gaps in knowledge while teaching that course," Scharman said. "I learned it working with those students. Dorothy used to point it out, and then I used to experience it. And I learned a lot from working with those kids as a result" (personal communication, August 15, 2014). Diana Rice, a former graduate instructor and professor of science education at Florida State University, said the course was her first opportunity to work with elementary education majors. "And that is the reason that I'm primarily still in elementary science education—because of that opportunity to work in Q200" (personal communication, August 20, 2014).

Gabel sought to fill the content and PCK gaps she had observed in her research by reforming her practice in the methods classroom. In the late 1980s, she submitted an NSF grant application to fund course development and further her research with preservice teachers. In 1990, she was awarded a $1.3 million grant from the NSF for Project QUEST (Quality University Elementary Science Teacher Preparation). Gabel shaped new courses at Indiana University from her research findings. Specific program goals for the grant included improving students' conceptual understanding of science, modifying their views of the nature of scientific inquiry, and validating the use of technology for science instruction. This NSF grant provided Gabel an opportunity to "take a broad-based approach to improving science education," including

- Upgrading IU science classes for elementary teachers, adding computers and hands-on experiences.
- Creating a new "capstone" course for elementary education majors who choose science as their area of concentration. The course,

called "Origins," intended to examine a science issue from the perspective of several disciplines.

- Training teachers in the Monroe County Community Schools to serve not only as teachers in their schools but also as mentors for IU elementary education students concentrating on science.
- Developing Saturday "science experiences" for Monroe County schoolchildren. (Hinnefeld, 1990)

The NSF grant allowed for further connections to be made between research and practice, specifically regarding the development and use of process skills as well as the development of key science concepts for future teachers. At the time, all elementary education students were required to take four 3-credit laboratory science courses: Introduction to Scientific Inquiry (Q200); a biological science course (either an introductory biology course or one specifically designed for elementary education majors); a physical science course (Q202, specifically designed for elementary education majors); and a geology course (students could choose one of three introductory courses offered in this discipline). As Sherwood explained, Gabel was able to create a bridge between scientists and educators. "She had enough chops in chemistry that they were willing to accept her as a scientist and that was important especially on that big grant because they were looking at ways to do better things in those early science courses for teachers" (personal communication, August 14, 2014).

This was now an opportunity for Gabel to develop elementary-education courses that included a large proportion of experimenting versus lecture and to use these new experiments to teach concepts in an inquiry manner. Other changes included a greater use of technology and an emphasis on teaching/learning using a constructivist approach. Within the Q200 course, important areas of focus included understanding of the macroscopic properties of matter and the continuity of concepts. The development of basic and integrated scientific process skills, such as graphing, problem solving, data interpretation, and communication skills, had been emphasized in her earlier effort for course improvement and was continuously incorporated into the course and corresponding text, *Introductory Science Skills,* published by Waveland Press (Gabel, 1993a). Preservice teachers taking the Q200 course were encouraged to use scientific reasoning and critical thinking to develop their understanding of science concepts. The overall objective allowed for more in-depth study of fewer topics so that concepts could be taught in a variety of contexts. In addition, teachers could stimulate learning by making the course content more related to what prospective teachers would need in teaching elementary science (Boone & Gabel, 1998). The course included these major concepts, among others:

- Matter is made up of elements, compounds, and mixtures.
- Matter has three states and can be changed from one state to another.
- Matter has mass and volume.
- Elements can exist in more than one form and combine to form compounds.
- Different elements have different properties.
- Elements and compounds can react chemically.
- Reactions can be speeded up and used to identify chemicals.
- Mass is conserved in chemical reactions.
- Chemical reactions can be represented by chemical equations.
- Diluting acids and bases changes their pH.
- Substances can be identified through their reactions.
- Substances have definite freezing points.
- Substances frequently have a crystalline structure.
- Crystals dissolve at different rates.
- The rate of dissolving depends on the temperature and surface area.
- Pure substances have definite boiling points.
- Solutions can be separated by boiling. (Gabel, 1993a)

For many graduate instructors, teaching the Q200 course has significantly impacted their science teaching philosophy and practices. Liang said that teaching Q200 profoundly enhanced her knowledge about scientific inquiry and continues to influence her own classroom practice today.

> Another relevant product of my teaching Q200 is the book on science education in elementary schools published in China several years later. In the book, my co-author Professor Huabin Chen (also a former doctoral student of Gabel's) and I introduced both theories and practices related to inquiry learning and teaching to Chinese elementary teachers. Some activities or examples were actually built upon our previous experiences with Q200. (personal communication, August 12, 2014)

To strengthen teacher development, Gabel acquired another NSF grant and created the SourceView media project, an American Chemical Society (ACS) resource on interactive teaching, from 1989 to 1994 (Gabel, ACS, & ChemSource, 1992). This project made it possible for preservice teachers to observe, along with their course instructor, a variety of chemistry teaching approaches on film. It also allowed an opportunity to make critical analyses of the video teaching segments that would often challenge preservice teachers' preconceived notions about good teaching.

As a science educator and research scholar, Gabel's scholarly work has always been closely connected with classroom teaching practices. She constantly used both research literature and her own empirical research findings to inform the practice. For instance, one of the curriculum development projects undertaken by Gabel was *Powerful Ideas in Physical Sciences* (PIPS), in collaboration with members of the American Association of Physics Teachers (AAPT). "While the book focused on main topics in physics for preservice elementary teachers, Gabel contributed one section on the nature of matter, and she decided to have it field tested in Q200," Liang said. Liang continued to explain the instructional model adopted in this section and how her dissertation study was to examine the effectiveness of this instructional approach:

> In this unit, Gabel adopted a conceptual change model of instruction by asking students to think about their ideas individually followed by the discussion of group's ideas, and then having the students test and reflect on their initial ideas through experimentation, post-lab reading and discussion. My dissertation research was to study the effectiveness of the instructional approach used in the PIPS.... The study involved six instructors; each taught two sections, one using the traditional method of instruction and the other using the PIPS approach. The research found that the students' scores on the paper-pencil unit test were not statistically different between the treatment and comparison groups. The selected student interview data, however, seemed to suggest that those in the PIPS group developed greater understanding of the targeted science concepts by the end of the unit study. Furthermore, the video analysis of classroom instruction revealed that the PIPS students were more actively engaged in thinking and learning during the classes. These research findings were subsequently used to inform the revision of curriculum and instruction. (personal communication, August 12, 2014)

Gabel was also concerned about the professional development of in-service teachers. She enjoyed getting to know teachers and learning about the kinds of things they were doing with their students. "She would get very excited if someone was doing something innovative," said Enid Zimmerman, a former IU colleague, art educator, and friend. "I think she saw teachers very positively" (personal communication, August 22, 2014). She participated in several summer programs for in-service teachers, notably at the University of Maryland, where in 1985–1986 she served as an instructor for graduate courses in problem solving in chemistry. "Her interaction with the graduate students versus teachers was probably different," Diana Rice said.

> She was very supportive and much more flexible. When we were in the program together in Maryland, all of the teachers there really liked Dorothy because she was the most supportive professor [in the program]. And I think that was really where her interest really was—with teachers. (personal communication, August 20, 2014)

In her NARST presidential address, Gabel made the following statement about the need for more action research by teachers in higher education:

> I feel that we need to make a greater effort to involve teachers in action research. Teachers already know much about teaching—more than many of us do. But many are waiting to be invited to participate in research studies in which they examine students' preconceptions, or effective teaching strategies. It is through joint research studies that science instruction in the schools will improve, and we need to make a great effort in this regard. (Gabel, 1995b)

SATURDAY SCIENCE

The QUEST grant also allowed for the implementation of summer science workshops for area teachers and the creation of "Saturday Science," affecting the education of countless K–12 teachers and their students. In describing the goals of Saturday Science, Gabel stated, "The purpose is to have children come away with a positive feeling about science. These activities are hands-on oriented and very experimental. Although they require a lot of work, they can be a lot of fun" (Wessel, 1990). The Saturday workshops with children afforded Gabel, her graduate students, and her undergraduate elementary preservice teachers an opportunity to observe students and analyze learning at particular developmental levels. Workshops were categorized by grade level, elementary through middle school, and met for 2.5 hours for six consecutive Saturdays. Because these workshops were an opportunity for Gabel to engage graduate student instructors in teaching practices, they elicited authentic questions based on graduate students' own experiences. Gabel modeled research question formation, research techniques, the bridge between theory and practice, and such through the Saturday workshop experiences rather than having graduate students learn skills in isolation from their use in practice.

For instance, in 1999 she reported an observation of misconceptions among 4th- to 6th-grade children regarding physical and chemical change. She stated, "Children in a Saturday Science program at Indiana University were asked to describe the difference, [and] many children included the concept of reversibility in their answer. Students said that physical changes were reversible and chemical changes were irreversible" (Gabel, 1999, p. 548). And in a 2001 study, Gabel, Monaghan, and MaKinster reported children's conceptions of burning after five hours of instruction within one of two Saturday workshops, Chemistry Around Us and Temperature, Heat, and Fire. Using interviews as well as a pre- and posttest multiple-choice instrument, graduate student instructors "practiced" their emerging skills in conducting research (Gabel, Monaghan, & MaKinster, 2001).

Elementary preservice teachers were also part of the instructional team that conducted the workshops. They usually assisted the instructors with activity setup and answered children's questions during the workshop. These workshops afforded both the graduate and undergraduate students an opportunity to relate theory to practice. Gabel's influence on teaching and teachers proved to be enormous. In her American Chemical Society Award nomination letter, Bretz (2006) wrote, "There can be no doubt that Dorothy's research has improved the teaching of thousands of teachers and consequently, the learning of tens of thousands of students."

SUPPORTING WOMEN IN SCIENCE EDUCATION

Recognition of women's equality and right to equal opportunities was a pressing issue in the field of science education throughout the era of Gabel's training and professional life, and it continues to be relevant today. Although women have certainly been represented in the classroom, the levels at which women have equal opportunities to influence science education research has changed over time (Champagne & Klopfer, 1977). Rossiter (1995) described the early 1970s as a time when women were encouraged to pursue degrees in science but found academic institutions to have a male-dominated structure. Typical positions open to women scientists of the time included that of research associates or volunteer assistants. This was a period in which women scientists could not even apply for a grant unless a fully appointed faculty member, often male, made the application for them. Nevertheless, women scientists made genuine contributions to their fields, grew in professional stature, and laid the foundation for the period after 1972, when real breakthroughs on the status of women scientists in America occurred. At the time Gabel was in her graduate studies, the notable women scientists included chemist Anna Jane Harrison, whose research assisted in the understanding of the structure of many organic molecules through spectroscopy and who eventually became the first woman to be elected president of the American Chemical Society (Shearer & Shearer, 1997). From 1973 to 1979, the growth in the number of women scientists on faculties across the nation had been minimal. However, given the long history of underutilization of women scientists in the academy, it is conceivable that without Executive Order 11246, which extended the clause of the 1964 Civil Rights Act prohibiting discrimination in employment (Title VII) to cover institutions of higher education, no change would have occurred whatsoever.

In a 1977 study (Champagne & Klopfer, 1977) that reviewed 60 volumes of the journal *Science Education* between the years 1916 and 1976, the representation of women in science education was considered minimal at best. According to the data, out of all 60 volumes, only 13% of the articles were written by women. And the percentages showed a decreasing trend from the earliest 10-year period to the latest (Champagne & Klopfer, 1977, p. 437). "Measured in the relation to the male authors in the profession's journal, women in science education appear to have sustained a slight loss in influence as the years have gone by" (Champagne & Klopfer, 1977, p. 437). Challenges such as these were common for women in the field of science education during the 1960s and 1970s. And, as Jane Kahle described, in the early 1970s there were only a few women within national associations to act as role models. "We were two of only six women at the NARST annual meeting," she said, "but the fact that there were only six, it's mind-boggling now to think...when now it has to be over half women in NARST" (personal communication, September 2, 2014). Kahle and Gabel shared their frustrations "about being one of the few women of that era, salaries, and [to] anyone who would care to listen, we would talk to them" (personal communication, September 2014). Nonetheless, Gabel was, as Kahle described, "as liberated as any of us," and while "Purdue was a conservative campus" (personal communication, September 2, 2014), she thrived. Hired by Indiana University (with a starting salary of $12,500) as an assistant professor of science education, Gabel spent her career, 1974–2004, at IU. As Jane Kahle explained, that salary was not unexpected: "First of all she was a single woman and a [Catholic] sister, so she was primed to be underpaid. I started Purdue at $13,500, so it wasn't that far off! [laughing]" (personal communication, September 2, 2014).

Gabel had received small grants from the Spencer Foundation and a larger NSF grant, $52,000, awarded in 1979, to study high school chemistry students' memory structures relevant to problem solving (Gabel & Sherwood, 1983). As Sherwood explained,

> Mary Budd Rowe had come to campus and given a talk; she was also a NSF program officer. And Mary was always interested in getting more women represented in the science education research and she really encouraged Dorothy to submit and that was the first grant. (personal communication, August 2014)

But even with Mary's support, Gabel had to take a stand. Kahle explained,

> One thing that I remember most impressed me about Dorothy is that she had put in a proposal to NSF and she felt she got an unfair or biased review. And, by gosh, she took NSF on, and by the second review she got funded. I was just so impressed that anyone would do that. It was very unusual. (personal communication, September 2, 2014)

Others recognized Gabel's accomplishments and ability to stand up for herself. "Dorothy was very religious, very principled," said Joanne Peng, a former colleague,

> but yet at the same time I could see her not being shy about promoting herself. In academia, once you get a grant you are likely to get a second or third one and so on. She acted more like a man in academia. She was obviously very talented and became really accomplished in her efforts. And, possibly, coming from the religious order she was very disciplined. (personal communication, August 25, 2014)

The IU School of Education had few women on the faculty in the 1970s, but one colleague, Enid Zimmerman, noted that graduate women and faculty would often get together. Gabel was active in supporting women, especially women in science. Zimmerman explained, "We had a lot of the same interests around the same things—democracy, liberalism, and hearing people's voices from those who were being downtrodden.... She was someone who really, I think, helped women to succeed who were in education" (personal communication, August 22, 2014).

Women faculty slowly became more organized about mentoring young faculty. In the early years of Gabel's career at IU, there was a group of women faculty who were mentoring women graduate students on things such as how to find a job, how to respond to challenging classroom incidents or if male teachers exhibited gender bias, and how to be "professional" (personal communication, August 22, 2014). Zimmerman also explained,

> At one time, we had 135 faculty and at that time maybe only 15–20 were women and very few who were full professors. So in our generation there weren't that many [women faculty members] who started out in the beginning. I remember when Dorothy came, she did a lot of [teaching] large classes and a lot of running around observing teaching in the schools and doing that kind of time intensive work. She didn't get to teach graduate classes until much later on. Women didn't have an easy time. Dorothy would work 24/7 a lot. (personal communication, August 22, 2014)

Gabel had to overcome many challenges early in her career. Being a young female faculty member with the need to gain, through hard work, some kind of acceptance from the older faculty wasn't easy. She was the only woman on the faculty in the science education department for many years. "Being a sister was another complication; not having a family and being different," Zimmerman said.

> It was a challenge to her as well. She could be very stubborn about things, and high-minded, and that kind of a personality was very good for perseverance and getting things done, but that's what makes a person who they are. Her dedication and [staying] focused on goals sometimes alienated others. (personal communication, August 22, 2014)

Gabel had difficult relationships at work. She was not included in social events involving faculty, Zimmerman said, and she and her colleagues differed on foundational beliefs.

> I think she wanted to hold everybody to her standards; that did not endear her to them. And some colleagues were more pragmatic rather than principled when it came to accomplishing their research agendas, and she was very much into theoretical as well as pragmatic issues. (personal communication, August 22, 2014)

By the early 1980s, the university hired more faculty to teach the larger number of classes needed as the Baby Boomers entered college. However, less than 10% of the faculty in the School of Education were women. A women's commission was organized by the school to encourage mentoring among female faculty and graduate students. Gabel was able to reach out and encourage other women faculty through the weekly commission meetings. Peng described her experience as a positive one:

> She was very gracious and I always appreciated her as a colleague and [her willingness] to reach out to me. As an international faculty member it seems that I need to have someone initiate an invitation to me first before I reciprocate. I really appreciated her and other faculty members at a time when they took me under their wing to acculturate me into the university environment. (personal communication, August 25, 2014)

Gabel's support of Peng led to a research collaboration. Gabel asked to work with her on an NSF research project, Peng helped with the statistical design and analysis. "We worked very well together and through this work I began to understand her research line and her productivity in her research" (personal communication, August 25, 2014).

Peng continued to work with Gabel on university committees. In the early 1990s, the Dean of the School of Education at IU wanted to commission a committee to look at the gender equity among faculty salaries. Gabel was a full professor at the time and she reviewed the associate professors; Peng, an associate professor, reviewed the full professors. "In one instance," Peng said,

> Dorothy was supportive of a faculty member that was trying to bring rigor to her teaching and Dorothy felt that those were virtues that we needed to support. I totally agreed with her on that point. So I feel that a lot of her decisions and handling her responsibilities was very principled. It was very intensive work, but in the end we were very happy with it. And others were happy with the analysis of the recommendations. (personal communication, August 25, 2014)

Gabel was committed to a life of service, and she served on numerous national, state, and university committees in the same year. As Sister Angelo Collins explained, "The whole concept of service is so essential to who we are as [Dominican] sisters. I don't know any other way to say it, it's just how we exist" (personal communication, August 20, 2014). Gabel continued to serve as a mentor for graduate students as well as faculty colleagues. Liang recalled that as part of her dissertation, Gabel directed her to read relevant literature and did mock interviews with her on how to probe student understanding of science concepts. "I had little experience in education research, so she kind of held my hand and walked me through it. I highly appreciated that" (personal communication August 12, 2014). Another graduate student, Diana Rice, explained, "She was a great role model in terms of how to do it [research]. And she gave me the opportunity to work with her on a project. And she directed me to the right research courses" (personal communication, August 20, 2014). Just as Dudley Herron had recruited and mentored Gabel through her graduate program, Gabel paid that forward by recruiting Rice for the IU doctoral program in 1986. "After I left that summer program in Maryland," Rice said,

> she had told a few of us that if we were interested in grad school we should let her know. She asked me if I were interested and I said yes. I really was, and so she helped review my graduate application and make sure it went through smoothly. So in that sense, you know, she recruited me. (personal communication, August 20, 2014)

Rice initially entered the program thinking that she would become a curriculum coordinator, but under Gabel's mentorship she developed the skills she needed to work at a university. "But the longer that I was there," Rice said, "I think in part because of the whole experience, but also because of Dorothy, I really went in the direction of being at a university" (personal communication, August 20, 2014). Dudley Herron trusted Gabel so much that when he left for Malaysia, he left Gabel in charge of his classes and with the responsibility to set up the new chemistry labs at Purdue. Gabel had a similar experience with Rice, giving her the responsibility for the Q200 course while she was serving at NSF in 1989. "She had already known me and she had seen me as a [veteran] teacher," Rice said,

> In fact, that first semester, she knew she was going [to NSF], so I taught one class and worked with her, so she was kind of preparing me. I was the coordinator while she was gone. She had her ideas about how things should be done. But she knew she could leave me. I did change a few things in the course while she was gone [laughing]. (personal communication, August 20, 2014)

Rice was a nontraditional graduate student in the sense that she was in her 30s and had a family while completing the program. It wasn't always easy to juggle family and graduate school responsibilities, but Rice felt that Gabel "really wanted women to be able to take the lead and be successful and I think she really realized that women often make compromises. She was very supportive of me but she also gave me a lot of independence" (personal communication, August 20, 2014). "Maybe she at least found something in our relationship that was more akin to what she had with Dudley because of the nature of how I met her, having a family, living not far from her." That allowed Rice to have a more personal relationship with Gabel. Rice said her children used to babysit for Gabel's Boston terrier, Buddy. "Buddy would have birthday parties and we would go over and celebrate, and all the grad students would be invited. It was rather interesting, because Buddy was her child. She had a few 'Buddys' [over the years]" (personal communication, August 20, 2014).

Gabel enjoyed the time she spent with Rice and her family. "She certainly enjoyed being around family and friends, and I think she always had sort of this nurturing circle around her," Kahle said (personal communication, September 2, 2104). Sherwood remembers bringing her son, who was born in 1977, into the office. "She just loved to see him. She really just softened that moment to see a baby. That showed a different side of her" (personal communication, August 14, 2014). Rice agreed: "She liked that 'family' kind of relationship. I think she had a real desire to have those kinds of relationships. You know because she didn't have much family, Carl was it. But you had to get beyond Dorothy's exterior, I think" (personal communication, August 20, 2014).

Gabel had high standards for her graduate students, and upon graduating they often took positions at research-intensive universities. "Today, many of her former students provide leadership to our field as influential researchers, science teacher educators, and policy makers" (R. Turner, email communication, August 28, 2014). "I think she took a great deal of joy, and it reflected on her personally, when we ended up at those types of schools," Rice said. "Especially the women, she took a great deal of pride that we ended up at the level that she thought we should be" (personal communication, August 20, 2014).

Gabel supported many international graduate students during her career. Liang remembered the first time she contacted Gabel was in 1993 when she applied for graduate school at IU.

> I had applied for multiple physics programs at other universities, but Dr. Gabel wrote me a very enthusiastic letter and told me that my academic record was excellent and she was willing to offer me a full graduate assistant scholarship to attend the program at IU. Deep in her heart, she was very kind, very straightforward, with little prejudice. She guided and nurtured the inter-

> national students from diverse cultures, which is probably one of her legacies among the IU faculty. (personal communication, August 12, 2014)

Liang's work with Gabel also led to a collaboration between IU faculty and faculty from China's Nanjing Normal University (NNU). Liang wholeheartedly recommended Gabel to the science education faculty members at NNU in China. Gabel was invited to NNU to give two talks on middle school science curriculum and instruction, and both were extremely well received. They also toured the nearby cities together afterwards (L. Liang, personal communication, August 12, 2014).

Accomplished chemistry educator Stacey Bretz met Gabel at a conference. "Her interest in my research and sincere comments of critique and suggestions were a highlight of that conference for me," Bretz said. Bretz recognized the significant mentoring that Gabel had undertaken regarding scholars in chemistry education.

> She has shared her intellectual insights with countless colleagues throughout her career. She has served on advisory boards, lending her support and reputation to ensure that fledgling, innovative, and sometimes risky projects will succeed. Dorothy Gabel has in one way or another influenced most every scholar in the field of chemistry education research today. (Bretz, 2006)

In 2000, Gabel was awarded the Perham Indiana Women of Achievement Award, given to Indiana women who enrich the lives of others through their accomplishments in a variety of fields. In this case, the award was for distinction in science education. The award, started in 1999, honors the memory of Bernadette H. Perham, who was an accomplished professor of mathematics and recognized for breaking down barriers to achievement for females in all disciplines. Gabel was certainly deserving of this award. She touched the lives of so many women in science education and was also an inspiration to women and men well outside her professional discipline.

BRIDGING SCIENCE AND SCIENCE EDUCATION

At a time when educational research was regarded as a "red-headed stepchild" and rarely trusted by scientists, policymakers, practitioners, and the public, Gabel's ability to bridge the education and the science community was of great value. As Dudley Herron explained, the field of science education was moving from its early practitioners-only focus to become a broader discipline that could also support legitimate research.

> You've got to understand that all of science education for its early life, everyone was practicing in that field, these were people who wanted to be teachers,

> and so they taught science. And then somebody told them what they told me: "Oh, you should go back and get your PhD and make contributions in these other areas." But their orientation is very much a service orientation, not a scholarly orientation. (personal communication, August 13, 2014)

In the 1960s, educational research was in its infancy and the "field of education as a subject for the kind of intellectual inquiry known as research" was just beginning (Smith, 1963, p. 200). A renewed focus on the science disciplines in the late 1960s led to a new source of criticism of education research. "In addition, though somewhat less noticed, there were self-conscious efforts to increase the disciplinary or academic rigor of many subfields of education research" (Lagemann, 2000, p. 165). This was certainly the case at Purdue University. Herron explained,

> Dorothy's PhD was in science education, because at that time I did not have graduate school status in chemistry. We wrestled for a long time with where does this animal belong? Because the nature of research is clearly rooted in education, in psychology, and cognitive science, and those are strange animals for chemists. There was a time when they didn't really trust each other. (personal communication, August 13, 2014)

Having already achieved two degrees in chemistry, Gabel was essentially living within the two cultures of science and education at Purdue. As Herron explained, Gabel was applying knowledge generated by psychologists and social scientists to problems of teaching and learning chemistry and communicating the results to chemists living well inside the world of science (Herron, 2008, p. 32). Herron described the time period as one of tumult.

> The gulf between science and humanities that [C. P.] Snow described almost 50 years ago may have narrowed, but it has not disappeared. Chemistry educators, who draw from both areas, are inevitably misunderstood in one culture or the other; like everyone else, they are occasionally subjected to criticism. Accept it. Learn from it. (Herron, 2008, pp. 29–31)

She certainly did. While at Indiana University, Gabel contributed substantially to both worlds. Problem solving is what chemists *do*, and Gabel had developed excellent problem-solving skills in her chemistry courses at Purdue. Gabel's collaboration with Herron was opportune. She was passionate about both the discipline and teaching.

The challenges faced by students learning chemistry in the classroom intrigued Gabel and led her to ask research questions that would determine the most effective strategies to teach problem solving, as well as to find strategies for building problem-solving skills. This became her life's work. When discussing science education reform on long car trips to visit her sisters on the East Coast, "Dorothy was quite sure that her work was the answer," said Sister Angelo Collins (personal communication, August 27, 2014).

Gabel was rigorous about her research standards and kept in touch with both the scholarship and classroom realities, said Herron (personal communication, August 13, 2014). "Gabel's knowledge and application of experimental designs appropriate for parametric statistical analysis is exceptional and to be envied," reported Susan Nurrenburn, a fellow chemistry educator, "and her work sets the standard for attention to experimental detail" (Nurrenburn, 2006). This ensured a certain level of respect from the scientists she would encounter as she worked to reform chemistry education. As a champion of the need for quality chemistry education research, Gabel was awarded the NARST Research Application Award on five occasions (1981, 1982, 1983, 1986, and 1987). The NARST award recognizes authors of the most outstanding paper, emphasizing classroom applications of research, presented at the annual meeting. In 1990 she shared her insights through NSTA's *What Research Says to the Science Teacher: Problem Solving* (Gabel, 1990a), and in 1994 she and Bunce contributed the chapter "Research on Problem Solving: Chemistry" in the *Handbook of Research on Science Teaching and Learning* published by Macmillan.

Gabel sought opportunities to interact with scientists. For instance, she acted as NSTA representative to the American Association for the Advancement of Science (AAAS) between 1985 and 1988 and continued working with AAAS as a delegate to Section Q (Education), between 1990 and 1992. Gabel also maintained her membership in the Indiana Academy of Science from 1975 to 1990. In addition, Gabel acted as a visiting professor in the Department of Chemistry at the University of Argentina in 1997. Staying connected to both chemistry *and* education, Herron said, was a significant contribution to science education (personal communication, August 13, 2014).

Gabel supported her graduate students with grant funding and encouraged them to collaborate closely with the IU scientists. Grad student Mike Svec recalled,

> My dissertation was on physics and how students read graphs and whether or not their reading of graphs contributed to their ideas about force and motion. That emerged out of the QUEST grant and working with Cathy Olmer, a scientist from the IU Physics Department. So the research really took place in the physics context, but in terms of the money, support, well, Dorothy's grant made that possible. (personal communication, August 14, 2014)

The best part, Svec said, was interacting with scientists who wanted to work with educators. "I think as a grad student that was coming from the sciences and new to education, having that connection with the scientists and getting that validation from both the scientists and the teacher education crew was important, and she did that well" (personal communication, August 14,

2014). Science education faculty view their work with graduate students in various ways. Some see the relationship as a mutual responsibility to move the student through the degree program successfully and/or a means to train research assistants using an apprenticeship model. Not unlike how graduate students in science are trained to become scientists, Gabel set up a community of practice that involved "ways of doing and approaching things that are shared to some significant extent among members" (Feldman, Divoll, & Rogan-Klyve, 2009, p. 446). This allowed Gabel and the students to have a certain level of mutual engagement in this joint enterprise of degree completion and researcher/practitioner training. Graduate students agreed that while working with Gabel they had increased their ability to ask researchable questions, make observations and collect data, understand the importance of controls, interpret data, communicate their results, and think independently. In essence, Gabel modeled the standard research practices in the field of chemistry education using clear, simple, researchable questions such as "Just why do students have such difficulty in solving chemistry problems?" and "Are some ways of teaching students to solve problems more effective than others?" To work out a research problem, she was prepared to work at it longer and harder than most people.

Gabel modeled the guidelines for publication, as well as the review process for journal articles, conference papers, and funding proposals. Scharmann stated, "I did take a seminar with Dorothy while I was there on how to review research, and we'd go through a paper and analyze the subjects, treatment, and the research design" (personal communication, August 15, 2014). Gabel also modeled the importance of inquiry and having a research focus or research agenda. Scharmann stated,

> She was very focused. If you are going to be balanced and have teaching, research, and service inform each other, you have to have a focal point. You know Dorothy had focal points in problem solving, in Piagetian work, and in the particulate nature of matter and how to communicate it. (personal communication, August 15, 2014)

She was also instrumental in bridging math and science disciplines, acting as the president of the School Science and Math Association (SSMA) between 1988 and 1990. A member since the early 1970s, Gabel was instrumental in supporting the work of Dr. Arthur White. "I was interested in integrating math and science in the classroom," said White (personal communication, September 22, 2014). When Gabel took a leave from IU to work at NSF in 1989, White met with Gabel, who was able to provide details on the best approach to use when seeking grant support from NSF. Gabel suggested that White write a proposal to hold a national conference on the subject of math and science integration. White explained,

> So I took that information back with me and, together with a colleague, Dr. Berlin, we wrote a proposal based upon her recommendations and we were funded in 1989, at $133,000, and were able to hold that conference. That was the beginning of the work on integrating math and science, and she helped us make that connection. The combination of these activities and explorations into integration resulted in over 150 publications and presentations worldwide and can be traced back to Dorothy Gabel's dedication to science education and the teaching and learning of science. (personal communication, September 22, 2014)

Gabel said the SSMA provided an opportunity for mathematics and science teachers at every level to meet one another, to exchange ideas, and to become aware of the goals of each discipline. The integration of mathematics and science that capitalizes on the overlap between the disciplines, Gabel said, would make both science and mathematics relevant to today's students (Gabel,1990b).

In 1999, Gabel and Nurrenburn produced the *Indiana Core 40 Assessment: Chemistry*, an assessment tool that emphasized a problem-solving focus. Even teachers who "teach to the test" were becoming aware of the research on effective instruction. Gabel had recognized the rapid increase and development of information technology and its potential impact on student learning using techniques such as 3-D imaging. "Chemistry education researchers need to think of the future and move forward in the areas that will be of greatest importance in the 21st century," she wrote. "Concurrently, there is a need to focus on ways to incorporate current research findings into the teaching and learning of chemistry today" (Gabel, 1999, p. 553).

Further evidence of Gabel's impact on science education through research comes from the numerous invited presentations she conducted in other countries—more than 30 presentations in 13 countries (D. Herron, personal communication, August 13, 2014). Her international reputation led to "numerous invitations as plenary and keynote speaker, including at such conferences as the International Conference on Chemistry Education, the Mexican Congress on Chemical Education, and the New Millenium Conference in Haifa, Israel. She also gave seminars in Germany, Taiwan, Australia, Sweden" (Bretz, 2006). Other invitations to share her expertise came from Poland, Argentina, Canada, England, Italy, and China. "Such invitations," Herron said, "certainly suggested that science educators around the world viewed Gabel's work as significant" (personal communication, August 13, 2014). Even after retirement, Gabel was awarded a Fulbright Senior Specialist Award in 2004 to teach at the National University of Quilmes (Argentina), where she conducted workshops for university faculty and high school science teachers about the implications of her research for the classroom" (Bretz, 2006).

"She made contributions in the area of bridging the gap and an understanding of the secular world and her religious world," Herron said.

> Dorothy was a person where both science and religion were extremely influential in her life, and she cared deeply about both of them. And so by the way she lived out her life, she was making a contribution that doesn't show up in her publications. (personal communication, August 14, 2014)

Dorothy Gabel is a unique and contributory scientist and science educator and her influence on the field of science education will be evidenced in the future as many of the colleagues and former students she has collaborated with will carry on her legacy.

ACKNOWLEDGMENTS

We wish to acknowledge the support of our colleagues, Dudley Herron, Enid Zimmerman, Ling L. Liang, Diana Rice, Larry Scharman, Mike Svec, Bob Sherwood, John Staver, Joanne Peng, Sr. Angelo Collins, Sr. Joan Brichacek, Arthur White, Carl Gabel, Jane Butler-Kahle, and others, in their willingness to be interviewed for this chapter on Dorothy's life and career. Thank you all for your dedication and your enthusiasm for fulfilling her vision for a brighter future.

REFERENCES

Abbott, D. (1984). *The biographical dictionary of scientists.* New York, NY: Bedrick.

Bodner, G., & Herron, J. D. (2002). *Chemical education: Towards research-based practice* (J. Gilbert et al., Eds.). Dordrecht, The Netherlands: Kluwer Academic.

Boone, W. J., & Gabel, D. L. (1998, January 1). Effectiveness of a model teacher preparation program for the elementary level. *Journal of Science Teacher Education, 9*(1), 63–84.

Bretz, S. L. (2006). *Nomination Letter.* Letter to ACS Award Selection Committee.

Cardellini, L. (2001). An interview with Dorothy L. Gabel [in Italian]. *La Chimica nella Scuola,* XXIII, 165–168. Retrieved August 8, 2014, from https://www.soc.chim.it/sites/default/files/users/div_didattica/PDF/2001-5.pdf

Cardellini, L. (2005). An interview with Dorothy L. Gabel. *Journal of Science Education, 6*(1), 9–11.

Champagne, A. B., & Klopfer, L. E. (1977, October 1). A sixty-year perspective on three issues in science teaching: I. Whose ideas are dominant? II. Representation of women. III. Reflective thinking and problem solving. *Science Education, 61*(4), 431–452.

CHED. (2008, Fall). *CHE newsletter, fall 2008.* Retrieved August 8, 2014, from http://www.divched.org/sites/default/files/2008fall.pdf

DeBoer, G. E. (2000, August 1). Scientific literacy: Another look at its historical and contemporary meanings and its relationship to science education reform. *Journal of Research in Science Teaching, 37*(6), 582–601.

Dominican University. (2014). *University mission and history.* Retrieved August 25, 2014, from http://www.dom.edu/about/mission

Dorin, H., Demmin, P. E., & Gabel, D. L. (1989). *Chemistry: The study of matter.* Needham, MA: Prentice Hall.

Feldman, A., Divoll, K., & Rogan-Klyve, A. (2009, April 1). Research education of new scientists: Implications for science teacher education. *Journal of Research in Science Teaching, 46*(4), 442–459.

Gabel, D. L. (1981). *Facilitating problem solving in high school chemistry.* Bloomington: Indiana University, School of Education. (ERIC Document Reproduction Service No. ED 210 192)

Gabel, D. L. (1983, October 1). High school chemistry textbooks: Form and function—A symposium: What high school chemistry texts do well and what they do poorly. *Journal of Chemical Education, 60*(10), 893–895.

Gabel, D. L. (1985). Chemistry for gifted children in the intermediate grades. *Journal of Chemical Education, 62*(5), 703–704.

Gabel, D. L. (1989a, September 1). *Chemistry resource materials.* Needham, MA: Prentice Hall.

Gabel, D. L. (1989b) Let us go back to nature study. *Journal of Chemical Education, 66*(9), 727–729.

Gabel, D. L. (1990a). What research says to the science teacher: Problem solving. *NARST.* Retrieved August 14, 2014, from https://www.narst.org/publications/research.cfm

Gabel, D. L. (1990b). Message from SSMA president Gabel. *School Science and Mathematics, 90*(1), 1.

Gabel, D. L. (1993a). *Introductory science skills.* Prospect Heights, IL: Waveland.

Gabel, D. L. (1993b, March 1). Use of the particle nature of matter in developing conceptual understanding. *Journal of Chemical Education, 70*(3), 193–194.

Gabel, D. L. (1995a). Science. In G. Cawelti & Alliance for Curriculum Reform (Eds.), *Handbook of research on improving student achievement.* Arlington, VA: Educational Research Service.

Gabel, D. L. (1995b, April 24). *NARST presidential address.* Address conducted at NARST annual conference, San Francisco, CA.

Gabel, D. L. (1998). Complexity of chemistry and implications for teaching. In B. J. Fraser & K. Tobin (Eds.), *International handbook of science education research.* Norwell, MA: Kluwer Academic.

Gabel, D. L. (1999, April 1). Improving teaching and learning through chemistry education research: A look to the future. *Journal of Chemical Education, 76*(4), 548–554.

Gabel, D. L., American Chemical Society, & ChemSource. (1992). *SourceView.* Washington, D.C: American Chemical Society.

Gabel, D. L., & Bunce, D. (1994). Research on problem solving: Chemistry. In D. L. Gabel & National Science Teachers Association (Eds.), *Handbook of research on science teaching and learning* (pp. 301–326). New York, NY: Macmillan.

Gabel, D. L., & Enochs, L. (1984). Preservice elementary teachers' conception of volume. *School Science and Mathematics, 84*(8), 670–680.

Gabel, D., & Herron, J. D. (1977). The effects of grouping and pacing on learning rate, attitude, and retention in ISCS classrooms. *Journal of Research in Science Teaching, 14*(5), 385–399.

Gabel, D. L., Monaghan, D., & MaKinster, J. (2001, December). Changing children's conceptions of burning. *School Science and Mathematics, 101*(8), 439–451.

Gabel, D. L., & National Science Teachers Association. (1994). *Handbook of research on science teaching and learning.* New York, NY: Macmillan.

Gabel, D. L., & Samuel, K. V. (1986, February 1). High school students' ability to solve molarity problems and their analog counterparts. *Journal of Research in Science Teaching, 23*(2), 165–176.

Gabel, D. L., & Samuel, K. V. (1987). Understanding the particle nature of matter. *Journal of Chemical Education, 64,* 696–697.

Gabel, D., Samuel. K. V., Helgeson, S., McGuire, S., Novak, J., & Butzow, J. (1987). Science education research interests of elementary teachers. *Journal of Research in Science Teaching, 24*(7), 659–677.

Gabel, D. L., & Sherwood, R. D. (1983, February 1). Facilitating problem solving in high school chemistry. *Journal of Research in Science Teaching, 20*(2), 163–177.

Gabel, D. L., & Sherwood, R. D. (1984, November 1). Analyzing difficulties with mole-concept tasks by using familiar analog tasks. *Journal of Research in Science Teaching, 21*(8), 843–851.

Gabel, D. L., Sherwood, R. D., & Enochs, L. (1984, February 1). Problem-solving skills of high school chemistry students. *Journal of Research in Science Teaching, 21*(2), 221–233.

Gabel, L. D., & Stucky, K. J. (2008). Prior knowledge of chemistry students: Chemistry K–8. In S. L. Bletz (Ed.), *Chemistry in the national science education standards: Models for meaningful learning in the high school chemistry classroom* (pp. 129–136). Washington, DC: American Chemical Society.

Gilbert, J. K., & Treagust, D. (Eds.). (2009). *Multiple representations in chemical education.* The Netherlands: Springer.

Haber, L. (1979). *Women pioneers of science.* New York, NY: Harcourt Brace Jovanovich.

Herron, J. D. (2008, January 1). Advice to my intellectual grandchildren. *Journal of Chemical Education, 85*(1), 24.

Herron, J. D. (2009a, August 29). *Sr. Gibson.* Letter to Sr. Mary Joan Gibson.

Herron, J. D. (2009b, September 24). *Award remarks.* Outstanding alumni awards ceremony, Purdue University, Lafayette, IN.

Hinnefeld, S. (1990, September 20). IU project aims at better science teaching. *The Herald-Times* [Bloomington, Indiana]. Retrieved August 12, 2014, from http://ww.heraldtimesonline.com//stories/1990/09/20/archive.19900920.f0ec5a6.sto?code=9e569e42-48ce-11e4-8884-1b3118ce8270

Howe, A. C., & Early, M. (1979). Reading and reasoning in ISCS classes. *Science Education, 63,* 15–23.

Indiana University. (1981, March 4). *Teacher criticizes emphasis on memorizing facts.* [News release].

Lagemann, E. C. (2000). *An elusive science: The troubling history of education research.* Chicago, IL: University of Chicago Press.

Nurrenburn, S. (2006, October). *Nomination letter.* Letter to ACS Award Selection Committee.

Rossiter, M. W. (1995). *Women scientists in America: Before affirmative action, 1940–1972.* Baltimore, MD: Johns Hopkins University Press.

Shearer, B. F., & Shearer, B. S. (1997). *Notable women in the physical sciences: A biographical dictionary.* Westport, CT: Greenwood.

Smith, H. A. (1963, September 1). Educational research related to science instruction for the elementary and junior high school: A review and commentary. *Journal of Research in Science Teaching, 1*(3), 199–225.

Staver, J. R., & Gabel, D. L. (1979, November 1). The development and construct validation of a group-administered test of formal thought. *Journal of Research in Science Teaching, 16*(6), 535–544.

Wessel, K. (1990, October 4). Professor wants to make science fun. *Indiana Daily Student,* p. 2.

CHAPTER 2

COLLABORATION, COMPILATIONS, AND SYNTHESES TO GROW SCIENCE EDUCATION

The Work of Stan Helgeson

Peter Rillero
Arizona State University

INTRODUCTION

The impact of a science educator is made in diverse ways through scholarship, service, and teaching. There are two key characteristics of truly important contributions: the work helps to grow the field of science education and the work helps inform the practice of science education. The work of Stan Helgeson was monumental in both these areas.

To his colleagues in science education, he was always Stan, an amiable gentleman who made others feel comfortable. He was a robust catalyst for bringing people together to improve science education. To me he will always be Dr. Helgeson, my professor and major advisor. A friendly, humble

Going Back for Our Future II, pages 41–56

man with a passion for teaching, he was always there to help. This chapter seeks to describe key contributions Helgeson made to science education.

BUILDING ACCESS TO SCIENCE EDUCATION

Science educators seeking information or doing literature reviews from 1966 to 1999 have almost certainly used the Education Resources Information Center database, more commonly known as ERIC, the world's largest educational database. Like other science educators, I used it to access an ocean of science education materials in my career; and I was also fortunate to have the experience of being the science education analyst for ERIC, reading articles and reports, writing their abstracts, and assigning controlled vocabulary to their indexing to enable efficient searches.

Prior to 1966, this resource for science education did not exist but the need for it was apparent. Science education is a relatively young discipline (Yager, 1980). Science educator researchers wanted their field to grow; to do so required access to what others were doing so they could be informed and build upon established work. In discussing the limited time many university faculty members had for research, a summary of a 1966 conference for the improvement of science education research contained the following:

> The problem is further complicated by the inaccessibility of previous research reports; absence of an organization or agency to provide adequate assistance in making literature searches adds to present limitations. A Coordinating Center could provide such services. It would enable the researcher to function more effectively and to devote more time to research design, analysis, and communication of results. (Richardson & Howe, 1966, p. 115)

The Center for Science Education

The ERIC Clearinghouse on Science Education was established at The Ohio State University in 1966 with a grant of $122,031 (Trester, 1979). Stan Helgeson was in charge of science education from 1969 to 1991 at this ERIC Clearinghouse, which through a few name changes became the ERIC Clearinghouse for Science, Mathematics, and Environmental Education (ERIC/CSMEE). Helgeson was responsible for deciding which materials would go into the database, supervising analysts, and reviewing their work. In this role, the work of science educators was made available to people around the world. From conference papers to evaluation reports, the materials were made available in the then state-of-the art technology of microfiche—index card-sized stiff plastic sheets that contained images of the papers viewable through lighted readers. It is difficult to imagine many electronic technologies that can survive in an unchanged form for 30+ years, but microfiche did. It presented many resources, in many places, to many people concerned with education. The 1996 *Directory of ERIC Resource Collections* (Taheri, Pearce, & Boston, 1996) lists 1,029 organizations, which were mainly univerisities, that provided access to ERIC microfiche collections.

The science education analysts also wrote abstracts for the main science education journals, and these journals were available at university libraries. Before desktop computers made an impact in the 1980s, researchers used two books for finding science education work that was indexed in ERIC. The *Resources in Education* (RIE) book contained research reports and projects and the *Current Index to Journals in Education* contained abstracts for over 700 journals. These books pointed the way to the resources, and university libraries supplied the actual resources.

With the Internet not yet developed, ERIC was a very important method to distribute work. In journals, books, and papers, the reference lists, citations, and bibliographies would contain the ERIC Document Number for all work that was on ERIC microfiche (as show in Figure 2.1), in a similar way to reference lists today have URLs for specific work. Researchers who had their materials accepted by ERIC and put on microfiche had the potential for a wider audience. It was not standard practice, but many authors put any ERIC-accepted materials in the published section of the curriculum vitae.

Growing the Lingo

The development of a field requires the advance of the terminology used in the field. Communication is enhanced when a term represents the same or similar concepts for researchers. Part of Helgeson's role at ERIC was to have the vocabulary of the abstracts evolve with the field of science education. New terms were first introduced as identifiers, after they proved to be useful they were elevated to descriptor status, such as occurred with the

Oregon Department of Education. (1990). *Concept/process-based science. Science Curriculum. Science curriculum concept paper #4.* Salem, OR (ERIC Document Reproduction Services No. ED 326 421).

Figure 2.1 The old accepted APA reference method showing how a document could be accessed through ERIC.

term “problem based learning.” Awareness of these terms, accessible from the ERIC Thesaurus, were important for efficient accessing of research and promoted the growth of research areas. Federal funding agencies would also tie programs into the descriptors for easier access (e.g., see Educational Programs that Work, 1995).

To help provide clarity, descriptors include “scope notes,” “a brief statement of the intended usage of a Descriptor. It may be used to clarify an ambiguous term or to restrict the usage of a term” (Houston, 2001, p. xxvii). For example, “scientific enterprise” was established in 1966 as a descriptor with the following scope note: “Totality of systematic activity of the sciences as an institution involving processes, attitudes, ethics, and interrelationships of science with other institutions” (Houston, 2001, p. 288). The descriptors were also put into a hierarchy with broader terms and more specific terms. For instance, “demonstrations (science)” fell under the more general term of “demonstrations (educational).” Thus, with Helgeson at the helm of science education at ERIC/CSMEE, a controlled indexing language grew with science education research.

STAN: THE MAN

Stanley Leon Helgeson was born in 1939 and raised in rural Wisconsin. His first school consisted of two classrooms, one for grades 1 and 2 and the other for grades 3 to 8. There were a total of about 25 students in the school at any given time. Young Stan had two other students in his grade, they were both named Donna Jean. In sixth grade, the Helgesons moved, and Stan graduated from Black Earth High in 1952. It was a record-size graduating class of 27 students. Shortly after, he married his high school sweetheart, Sue.

In part for GI Bill Benefits regarding attending university and in part to avoid being drafted, Helgeson enlisted in the U.S. Army in 1954. He served at Army hospitals first as an admission clerk in Missouri and after that as a Medical Records Specialist in Germany. After an honorable discharge in 1956, he continued as a reservist for 18 months and earned the rank of Seargent.

The GI Bill did pay for his education, allowing Helgeson to earn his BS in secondary education (biology) in 1960 from the University of Wisconsin, and he started teaching science at Chippewa Falls Junior High School. Great concern was being expressed in the United States at that time about the Russian space program. The NSF initiated an Academic Year Institute for teachers to go back to college and earn additional credits. Helgeson applied and was accepted to Arizona State University, where he earned a Master of Natural Science degree in June 1964. At about the same time, the University of Wisconsin offered a second-year institute for people to

Figure 2.2 Stan Helgeson.

become science education supervisors. Helgeson entered, and this was entrée into their PhD program where his major advisor was Milton Pella. Graduating in 1967, Helgeson worked for one year at University of Toledo before starting the rest of his career at Ohio State Univeristy in 1968. In 1992, he transitioned from ERIC/CSMEE to the National Center for Science Teaching and Learning, which was a new project at Ohio State, where he was Director of the Focus Area 4: New Technologies. Helgeson retired from Ohio State with emeritus status in 1995 and passed away in 2013.

PROMOTING THE RESEARCH

Helgeson embraced the goal of getting people to use the research in science education. This audience was both domestic and international, researcher and practitioner. From Anaheim to Taiwan, Helgeson traveled from place to place informing and exciting people about the science education contents of ERIC and science education research. At science education conferences, ERIC had a table to help educate new researchers about how to find science education resources and to help remind established researchers about submitting their work into ERIC. Helgeson was an active member of NARST, NSTA, AETS, and the School Science and Mathematics Association. For example, at NARST he was research coordinator for 6 years, president for 2 years, and on the Executive Board from 1971 to 1977 and from 1979 to 1983. In coordination with NSTA, AETS, and NARST,

Helgeson facilitated the development of several book series that helped shape the field of science education.

Investigations Series

The *Investigations* series was an impressive long-term effort to improve and promote science research. The first volume was published in 1974 (Helgeson & Blosser, 1974) and the last was in 1990, with several issues of each volume produced per year. About 8 to 10 science education research journal articles were selected for each issue, and in their own words, another science education researcher described what the rational, methods, and conclusions were for the article. These syntheses presented the research in a very clear way, but perhaps most interesting was when reviewers went beyond the article and input their thoughts, analysis, review, and critique in the section called the "Abstractor's Analysis."

As an example, Bredderman's 1974 study of science experiences and the ability to control variables was reviewed by Lawson (1977), where the analysis moved into why the SAPA program wasn't effective in having students control variables, while another study showed what the SCIS program was. Also fascinating to read is Baker's 1988 review of DeBoer's 1984 article on gender effects, with Baker passionately promoting more research in gender issues and science education.

The articles in the *Investigations Series* were useful and interesting reading for current and future researchers. While many might consider the peer-review process to produce only high-quality articles, it was useful to have another researcher go through and explain someone else's research in their own and different way and then move to a critique pointing out the positives and negatives of the research. Some reviews were critical of the published research, such as Ziegler's (1975) review of Bybee's "The Teacher I Like Best" article, which added interest to the series. As with other ERIC/CSMEE publications, the *Investigations* books could be purchased for a low cost from the ERIC Clearinghouse, for example, Volume 2, Issue 2 was sold for $1.75, or it could be accessed for free from a microfiche collection.

The AETS Yearbooks

The AETS Yearbooks was another important series that fell under the auspices of Helgeson at ERIC/CSMEE. The first yearbook was a review of teacher education practices conducted by Balzar, Evans, and Blosser (1973). The second yearbook (*Science Teacher Education 1974: Issues and Positions*) changed to have a focus on a few topics, with chapters written by different individuals (Fox, 1975). For each chapter, there was another shorter chapter that was a reaction to that chapter. By 1978, the yearbook changed to be chapters without published reactions. The last yearbook of the series was published in 1995 (Otto, 1995). The yearbooks made great contributions to science teacher education.

The Summary of Research Series

The annual books summarizing research in science education ran throughout Helgeson's leadership in science education. These were useful descriptions and analyses of the state of science education. In addition to being separate books, for many of the years, the annual book was also published as an entire journal issue of *Science Education*, which effectively increased dissemination. For example, Gabel, Kagan, and Sherwood's (1979) *A Summary of Research in Science Education—1978* was published in book form by ERIC/CSMEE, where it later became freely available on microfiche. Later it became the *Science Education* journal edition (Gabel, Kagan, & Sherwood, 1980).

Starting for the period 1965 to 1967, the reports were published as three books for the levels elementary, secondary, and postsecondary. In a 1968–1969 summary of research, Helgeson and Blosser (1971) described the reports in the introduction as follows:

> Research Reviews are being issued to analyze and synthesize research related to the teaching and learning of science completed during a two-year period of time. These reviews are organized into three publications for each two-year cycle according to school levels—elementary school science, secondary school science, and college science.
>
> The publications are developed in cooperation with the National Association for Research in Science Teaching. Appointed NARST committees work with staff of the ERIC Center for Science Education to evaluate, review, analyze, and report research results. It is hoped that these reviews will provide research information for development personnel, ideas for future research, and an indication of trends in research in science education. (p. 2)

In the same 1968 to 1969 Summary of Research, Welch described where the research came from as follows:

> Of particular interest is the relative contribution to the total research effort made by the various types of activities. More than a third of the documents are dissertations, while one-sixth are nonjournal articles. This suggests that more than half of the total science education research effort is either an ini-

> tial research undertaking (dissertation) or a nonpublished report. (Welch, 1971, p. 2)

In 1970 the reports were all contained in one book and the review was for a 1-year period (Trowbridge, Anderson, & Rahn, 1972). In this book, the research in science education at the college, secondary, and elementary levels were all described separately. By the 1972 yearbook, the grade levels had merged, and the book was organized by research topics, including evaluation and testing, curriculum studies, computer-assisted instruction, attitudes toward science and science values, Bloom's taxonomy, and interaction analysis (Novak, 1973).

The evolution of the series reflected Helgeson's leadership and his vision that research in science education needs to be shared with a broad community and should be used to improve student learning. While establishing traditions was good, change was necessary in the rapidly developing field of science education. Gabel et al. described the goals of the 1978 yearbook as follows:

> The purposes of this review are virtually the same as those expressed by Rowe and DeTure's *A Summary of Research in Science Education*. These are (1) to portray the state of knowledge in science education, (2) to describe any existing trends, (3) to identify areas that need to be researched, and (4) to provide tentative answers to persistent problems, if any seem to emerge from the research. (1979, p. 4)

For this work and others, Helgeson put the right people in place and gave them resources to succeed. As one author noted, "The majority of studies in this review were identified by Dr. Stanley Helgeson and his staff at the Ohio State University who supplied us with abstracts of articles from major journals, dissertations, papers presented at meetings, and other ERIC holdings" (Gabel et al., 1979, p. 5).

NARST Collections of Detailed Abstracts

Helgeson took it upon himself and his team to collect, edit, and organize the detailed abstracts at NARST conferences and to publish these in ERIC. He also actively solicited full conference papers and release forms from authors for potential inclusion into ERIC. For NARST, this work started in earnest in 1969 with the collection of 44 abstracts from the 42nd Annual Meeting (ERIC Information Analysis Center for Science Education, 1969). Helgeson did the compilation until 1977, and after that colleagues, Art White and Pat Blosser participated, with the last compilation of this series published in ERIC in 1990 (Blosser & Helgeson, 1990). Today, where electronic documents are easily shared via the Internet, such work isn't needed.

But in the span of 1969 to 1990, these published documents helped the science education community access research, share work, and grow many subdomains within science education research.

Research and the Science Teacher

Helgeson had the goal of making research available, and an audience he was particularly interested in was classroom teachers. A culmination of this passion was the jointly produced ERIC/CSMEE and NSTA publication series *What Research Says to the Science Teacher.* This was produced annually for 5 years, with Mary Bud Rowe serving as editor for the first 2 years. In the second volume, she wrote,

> So much argument, indecision, and thrashing about in education goes on unimpeded by knowledge of research! No doubt this is true in part because teachers and other practitioners rarely have time to ferret out studies which, taken together, might form a basis for action or provide clues to the solution of practical problems. (Rowe, 1979, p. i)

The mission of this series was clear: to bring research to classroom teachers. Helgeson also used his position to point teachers toward useful science education materials for the classroom with his syntheses of 15 years of ERIC submissions (Helgeson, Blosser, & Howe, 1982). Although not part of this ERIC role, Helgeson served as an editor on two editions of *Theory Into Practice* (Gilliom, Helgeson, & Zuga 1991, 1992), which focused on science, technology, and society.

Emerging Environmental Education

While science education is a relatively new field of study, environmental education is even newer. The year 1970 stands out as the most prominent start-up year, with the first year of implementation of the landmark National Environmental Policy Act (1969), the first Earth Day, and a speech by President Nixon to Congress, where he stated,

> It is also vital that our entire society develop a new understanding and a new awareness of man's relation to his environment—what might be called "environmental literacy." This will require the development and teaching of environmental concepts at every point in the education process. (Nixon, 1970, p. vii)

In 1971, the National Association for Environmental Education was founded (McCrea, 2006). Helgeson was fast in supporting this new field. A series of publications helped inform environmental educators and environmental research. These included (a) *Review of Environmental Education for Elementary and Secondary School Teachers* (Helgeson, Helburn, Wiley, & Blosser, 1971); (b) the chapter "The Role of ERIC in Environmental Education" (Helgeson, 1971); and (c) *Environmental Programs and Materials* (Helgeson, Blosser, Howe, Helburn, & Wiley, 1972).

In their *Review of Research in Environmental Education*, Roth and Helgeson (1972) stated,

> Little attention has been given to searching-out, organizing, and reporting research in the environmental education area. The purpose of this report is to: (1) identify known research pertaining to environmental education, (2) review critically the identified research, and (3) identify areas for further research. (p. 7)

SYNTHESIS REPORTS

Among the achievements in helping to communicate and develop the field of science education, Helgeson directly authored many publications. He wrote several ERIC Digests, which were short summaries of research that were published as research intensified. Helgeson's published keynote address "Trends and Issues in Science Education" to the International Symposium on Science Education in Taiwan shows Helgeson's strong understanding of science education and his ability to communicate (Helgeson, 1992).

Helgeson's most influential report, *The Status of Pre-College Science, Mathematics, and Social Science Education: 1955–1975* (Helgeson, Blosser, & Howe, 1977) was commissioned by the National Science Foundation. This 20-year period was radically different from what came before and foundational for the future of science education.

> The two decades, 1955–1975, are unparalleled in the degree of activity in science education. Millions of dollars were devoted to the cooperative involvement of scientists, educators, and learning theorists in the development of science curriculum materials. Extensive programs were conducted to upgrade and update the science content background of teachers and to train them in the use of the new curricula. (Helgeson, Blosser, & Howe, 1977, p. 1)

One of the oft-quoted findings of the report was that 80% of the elementary teachers and 90% of the intermediate-grade teachers based their science instruction upon a single textbook. This status report became a key document in science-education research and projects. It provided part of the foundation for the influential Project Synthesis (Barrow, 2006). Weiss puts this work into context as follows:

> In 1976, in order to get a better idea of the status of science, mathematics, and social studies education in the United States, the National Science Foundation (NSF) commissioned three large studies. These included: (1) a major review of the science, mathematics, and social studies education research literature, coordinated by Stanley Helgeson at The Ohio State University; (2) a national survey of teachers, principals, district, and state personnel, directed by Iris Weiss, then at the Research Triangle Institute; and (3) intensive case studies in 11 districts, coordinated by Robert Stake and Jack Easley at the University of Illinois. The results of these three studies, collectively known as "the NSF needs assessment," were disseminated widely and used extensively in program decision making. (Weiss, Pasley, Smith, Banilower, & Heck, 2003, p. 1)

Prior to becoming president of NARST, Helgeson was the Research Coordinator for this research organization for 6 years. Using his yearly summaries and syntheses, he published in the *Journal of Research in Science Teaching* (*JRST*) the article, "Priorities for Research in Science Education" (Helgeson, 1978). Some 6 years after Gene Glass coined the term "meta-analysis" (O'Rourke, 2007), Helgeson was part of a team that produced a meta-analysis of 68 research studies on science-teacher education practices associated with inquiry strategies (Sweitzer, Howe, Helgeson, & Blosser, 1982). For the *Handbook of Research on Science Teaching and Learning Project* (Gabel, 1994), Helgeson authored the chapter "Research on Problem Solving: Middle School" (Helgeson, 1994) and co-authored the chapter "Research on Assessment in Science" (Doran, Lawrenz, &, Helgeson, 1994).

SCHOLARSHIP

Helgeson was cordial and collegial, and also had the respect of science-education researchers because he was a scholar. Although he focused much of his energy on sharing and promoting the research of others, he was a strong researcher. He was a pioneer in investigations of elementary students' conceptual understandings of force (see Helgeson, 1968). With a team that included Gabel, Novak, and Butzow, original research was done and articles published in *JRST* on research interests of teachers (see Gabel et al., 1987; Gabel, Samuel, Helgeson, Novak, & Butzow, 1986). With Kumar, Helgeson explored and reviewed the potential of computers to affect science education and published several research articles (Kumar & Helgeson, 1996, 2000; Kumar, Helgeson, & White, 1994). A prominent achievement was being awarded the Outstanding Research for Practical Applications NARST award in 1986 and again in 1987.

David Kumar describes his perspectives on Helgeson:

> Dr. Helgeson has been perhaps the most influential person in my science education path. He was not only a great mentor but also a great colleague and a person of integrity. He made numerous contributions to science education, including research studies, reviews and monographs, and playing an active role leading the ERIC Clearinghouse for Science, Mathematics and Environmental Education in its early days. It should be noted that Dr. Helgeson numbered among a handful of "those who did the research" upon which the most influential science education reform document ever, the Project Synthesis, was developed. In my opinion Dr. Stanley L. Helgeson was one of the top five science education professors who have ever lived in the United States. (D. Kumar, personal communication, July 4, 2014)

CONCLUSION

Helgeson helped grow the field of science education through his scholarship, collaboration, and publications, making science and environmental education materials available to researchers, teachers, and policymakers. Some of the things Helgeson helped build are thriving, including NARST and the field of science education. Some past efforts, while important in their day, are not being produced today, as electronic communication facilitates sharing and government cuts increase. For many of us, ERIC is still the "go to" database for science education literature reviews. While other privately created databases exist, the education focus of ERIC continues to make it relevant. The ERIC Clearinghouses are no longer funded, and the important publications of ERIC/CSMEE are no longer produced. The annual reviews of research and the Investigation Series are missed, and perhaps this chapter will help motivate revival efforts.

The body of work of Stan Helgeson was pivotal in growing the field of science education. He made the work of science educators available to the world. His scholarship and syntheses provided foundations for many to build upon. Helgeson, a cordial and amiable man, pulled many science educators together and provided guidance on the production of publications to disseminate work and grow the field. His professional focus of making science education research available and relevant was a powerful force in the development of this field.

REFERENCES

Baker, D. R. (1988). Expanded abstract and analysis of DeBoer's "A Study of Gender Effects in the Science and Mathematics Course-taking Behavior of a Group of

Students who Graduated From College in the Late 1970's." In P. E. Blosser & S. L. Helgeson (Eds.), *Investigations in science education, 14*(4), 61–67. Columbus, OH: ERIC Clearinghouse for Science, Mathematics, and Environmental Education. Retrieved from http://files.eric.ed.gov/fulltext/ED303037.pdf

Balzer, L. A., Evans, T. P., & Blosser, P. E. (1973). *A review of research on teacher behavior relating to science education.* Columbus, OH: ERIC Information Analysis Center for Science, Mathematics, and Environmental Education. Retrieved from http://files.eric.ed.gov/fulltext/ED087638.pdf

Barrow, L. H. (2006). A brief history of inquiry: From Dewey to standards. *Journal of Science Teacher Education, 17,* 265–278

Blosser, P. E., & Helgeson, S. L. (Eds.). (1990, April 8–11). *Abstracts of presented papers. 63rd National Association for Research in Science Teaching Annual Conference, Atlanta, GA.* Columbus, OH: ERIC Clearinghouse for Science, Mathematics, and Environmental Education. Retrieved from http://files.eric.ed.gov/fulltext/ED335235.pdf

Doran, R. L., Lawrenz, F., & Helgeson, S. (1994). Research on assessment in science. In D. L. Gabel (Ed.), *Handbook of research on science teaching and learning* (pp. 388–442). New York, NY: Macmillan

Educational Programs that Work. (1995). *Index II: ERIC descriptors.* Retrieved from http://www2.ed.gov/pubs/EPTW/eptwind2.html

ERIC Information Analysis Center for Science Education. (1969). *Abstracts of presented papers. National Association for Research in Science Teaching 42nd annual meeting Columbus, OH.* ERIC Information Analysis Center for Science and Mathematics. Retrieved from http://files.eric.ed.gov/fulltext/027227.pdf

Fox, F. W. (1975). *Science teacher education 1974: Issues and positions. 1975 AETS yearbook.* Columbus, OH: ERIC Information Analysis Center for Science, Mathematics, and Environmental Education. Retrieved from http://files.eric.ed.gov/fulltext/ED099244.pdf

Gabel, D. L. (Ed.). (1994). *Handbook of research on science teaching and learning.* New York, NY: Macmillan.

Gabel, D. L., Kagan, M. H., & Sherwood, R. D. (1979). *A summary of research in science education—1978.* Columbus, OH: ERIC Information Analysis Center for Science, Mathematics, and Environmental Education. Retrieved from http://files.eric.ed.gov/fulltext/ED183365.pdf

Gabel, D., Kagan, M., & Sherwood, R. (1980). A summary of research in science education—1978. *Science Education, 64,* 429–568.

Gabel, D., Samuel, K. V., Helgeson, S., McGuire, S., Novak, J., & Butzow, J. (1987), Science education research interests of elementary teachers. *Journal of Research in Science Teaching, 24*(7), 659–677.

Gabel, D. L., Samuel, K. V., Helgeson, S., Novak, J., & Butzow, J. (1986). Research interests of secondary science teachers. *Journal of Research in Science Teaching, 23*(2), 145–163.

Gilliom, E. M., Helgeson, S. H., & Zuga, K. F. (Eds.). (1991). Special Issue: Science, technology, society: Challenges. *Theory into Practice, 30*(4).

Gilliom, E. M., Helgeson, S. H., & Zuga, K. F. (Eds.). (1992). Special Issue: Science, technology, society: Opportunities. *Theory into Practice, 31*(1).

Helgeson, S. L. (1968). *The relationships between concepts of force attained and maturity as indicated by grade levels.* Madison, WI: Center for Cognitive Learning. Retrieved from http://files.eric.ed.gov/fulltext/ED020910.pdf

Helgeson, S. L. (1971). The role of ERIC in environmental education. In R. S. Cook & G. T. O'Hearn (Eds.), *Processes for a quality education.* Green Bay: University of Wisconsin-Green Bay.

Helgeson, S. (1978). Priorities for research in science education: An introduction. *Journal of Research in Science Teaching, 15*(2), 95–97.

Helgeson, S. L. (1992). *Trends and issues in science education.* Keynote address to the International Symposium on Science Education, Taipei, Taiwan. Retrieved from http://files.eric.ed.gov/fulltext/ED357954.pdf

Helgeson, S. L. (1994). Research on problem solving in middle school. In D. L. Gabel (Ed.), *Handbook of research on science teaching and learning* (pp. 248–268). New York, NY: Macmillan.

Helgeson, S. L., & Blosser, P. E. (1971). Research reviews—Science. In W. Welch (Ed.), *A summary of research in science education for the years of 1968–69, secondary school level* (pp. 2–3). Columbus, OH: ERIC Information Analysis Center for Science, Mathematics, and Environmental Education. Retrieved from http://files.eric.ed.gov/fulltext/ED052051.pdf

Helgeson, S. L., & Blosser, P. E. (Eds.). (1974). *Investigations in science education, volume 1 number 1, Expanded abstracts and critical analyses of recent research.* Columbus, OH: ERIC Information Analysis Center for Science, Mathematics, and Environmental Education. Retrieved from http://files.eric.ed.gov/fulltext/ED103274.pdf

Helgeson, S. L., Blosser, P. B., & Howe, R. W. (1977). *The status of pre-college science, mathematics, and social science education: 1955–1975. Volume 1, Science education.* Columbus, OH: Center for Science and Mathematics Education, The Ohio State University. Retrieved from http://files.eric.ed.gov/fulltext/ED153876.pdf

Helgeson, S., Blosser, P. B., & Howe, R. W. (1982). *Especially for teachers: Selected documents on the teaching of science 1966–1981.* Columbus, OH: ERIC Clearinghouse for Science, Mathematics, and Environmental Education. Retrieved from http://files.eric.ed.gov/fulltext/ED247080.pdf

Helgeson, S. L., Blosser, P. B., Howe, R. W., Helburn, N., & Wiley, K. B. (1972). *Environmental education programs and materials.* Retrieved from http://files.eric.ed.gov/fulltext/ED062582.pdf

Helgeson, S. L., Helburn, N., Wiley, K. B., & Blosser, P. E. (1971). *A review of environmental education for elementary and secondary school teachers, volume 1 of 3. Final report.* Columbus, OH: The Ohio State University Research Foundation. Retrieved from http://files.eric.ed.gov/fulltext/ED059913.pdf

Houston, J. E. (2001). *Thesaurus of ERIC descriptors* (14th ed.). Westport, CT: Oryx.

Kumar, D. D., & Helgeson, S. L. (1996). Effect of computer interfaces on chemistry problem solving among various ethnic groups: A comparison of pen-point and powerbook computers. *Journal of Science Education and Technology, 5*(2), 155–160.

Kumar, D. D., & Helgeson, S. L. (2000). Effect of gender on computer-based chemistry problem solving: Early findings. *Electronic Journal of Science Education, 4*(4)

Retrieved from http://wolfweb.unr.edu/homepage/crowther/ejse/kumaretal.html

Kumar, D. D., Helgeson, S. L., & White, A. L. (1994). Computer technology-cognitive psychology interface and science performance assessment. *Educational Technology Research and Development, 42*(4), 6–16.

Kumar, D. D., White, A. L., & Helgeson, S. L. (1994). A study of the effect of HyperCard and pen-paper performance assessment methods on expert-novice chemistry problem solving. *Journal of Science Education and Technology, 3*(3), 187–200.

Lawson, A. (1977). Expanded abstract and analysis of Bredderman's "Elementary School Science Experience and the Ability to combine and Control Variables." In S. L. Helgeson & P. E. Blosser (Eds.), *Investigations in science education, volume 3 number 4, Expanded abstracts and critical analyses of recent research* (pp. 3–10). Columbus, OH: ERIC Information Analysis Center for Science, Mathematics, and Environmental Education. Retrieved from http://files.eric.ed.gov/fulltext/ED164326.pdf

McCrea, E. J. (2006). *The roots of environmental education: How the past supports the future.* Stevens Point: Environmental Education and Training Partnership, University of Wisconsin-Stevens Point. http://files.eric.ed.gov/fulltext/ED491084.pdf

National Environmental Policy Act. (1969). *The National Environmental Policy Act of 1969.* Retrieved from http://ceq.hss.doe.gov/nepa/regs/nepa/nepaeqia.htm

Nixon, R. (1970). *President's message to the Congress of the United States. In Environmental quality, the first annual report of the Council on Environmental Quality, together with the President's message to Congress.* Washington, DC: U.S. Government Printing Office. Retrieved from http://files.eric.ed.gov/fulltext/ED062 109.pdf

Novak, J. D. (1973). *A summary of research in science education—1972.* Columbus, OH: ERIC Information Analysis Center for Science, Mathematics, and Environmental Education. Retrieved from http://files.eric.ed.gov/fulltext/ED090055.pdf

O'Rourke, K. (2007). An historical perspective on meta-analysis: Dealing quantitatively with varying study results. *Journal of the Royal Society of Medicine, 100*(12), 579–582.

Otto, P. (Ed.). (1995). *Science education in the rural United States. Implications for the twenty-first century.* Columbus, OH: ERIC Clearinghouse for Science, Mathematics, and Environmental Education. Retrieved from http://files.eric.ed.gov/fulltext/ED390649.pdf

Richardson, J. S., & Howe, R. W. (1966). *The role of centers for science education in the production, demonstration, and dissemination of research.* Columbus, OH: The Ohio State University Research Department. Retrieved from http://files.eric.ed.gov/fulltext/ED013220.pdf

Roth, R. E., & Helgeson, S. L. (1972). *A review of research related to environmental education.* Columbus, OH: ERIC Information Analysis Center for Science, Mathematics, and Environmental Education. Retrieved from http://files.eric.ed.gov/fulltext/ED068359.pdf

Rowe, M. B. (1979). *What research says to the science teacher. Volume 2.* Washington, DC: National Science Teachers Association. Retrieved from http://files.eric.ed.gov/fulltext/ED166057.pdf

Sweitzer, G. L., Howe, R. W., Helgeson, S. L., & Blosser, P. B. (1982). *A meta-analysis of research on science teacher education practices associated with inquiry strategy.* Columbus, OH: ERIC Clearinghouse for Science, Mathematics, and Environmental Education. Retrieved from http://files.eric.ed.gov/fulltext/ED226991.pdf

Taheri, B. J., Pearce, E., & Boston, C. (1996). *Directory of ERIC resource collections 1996.* Washington, DC: US Department of Education. Retrieved from http://files.eric.ed.gov/fulltext/ED400831.pdf

Trester, D. J. (1979). *ERIC—The first fifteen years. A history of the Educational Resources Information Center.* Washington, DC: Educational Resources Information Center. Retrieved from http://files.eric.ed.gov/fulltext/ED195289.pdf

Trowbridge, L. W., Anderson, J. R., & Rahn, J. E. (1972). *A summary of research in elementary, secondary, and college levels of science education for 1970.* Columbus, OH: ERIC Information Analysis Center for Science, Mathematics, and Environmental Education. Retrieved from http://files.eric.ed.gov/fulltext/ED085219.pdf

Weiss, I. R, Pasley, J. D., Smith, P. S., Banilower, E. R., & Heck, D. J. (2003). *Looking inside the classroom: A study of K–12 mathematics and science education in the United States.* Chapel Hill, NC: Horizon Research. Retrieved from http://www.horizon-research.com/insidetheclassroom/reports/looking/complete.pdf

Welch, W. (1971). *A summary of research in science education for the years of 1968–69, secondary school level.* Columbus, OH: ERIC Information Analysis Center for Science, Mathematics, and Environmental Education. Retrieved from http://files.eric.ed.gov/fulltext/ED052051.pdf

Yager, R. E. (1980). *Crisis in science education. Technical report No. 21.* Iowa City, IA: Science Education Center. Retrieved from http://files.eric.ed.gov/fulltext/ED202664.pdf

Ziegler, A. (1975). Expanded abstract and analysis of Bybee's "The Teacher I Like Best: Perceptions of Advantaged, Average, and Disadvantaged Science Students." In S. L. Helgeson & P. E. Blosser, (Eds.), *Investigations in science education, volume 1 number 2, Expanded abstracts and critical analyses of recent research* (pp. 10–13). Columbus, OH: ERIC Information Analysis Center for Science, Mathematics, and Environmental Education. Retrieved from http://files.eric.ed.gov/fulltext/ED111637.pdf

CHAPTER 3

ROBERT K. JAMES

Kevin D. Finson
Bradley University

Janice D. Meyer
Pasadena (TX) Independent School District

Many biographical sketches seem to begin with when and where a particular person was born. However, perhaps it is fitting to begin with when and where Bob James's life among us ended. It was on December 28, 2012 in Decatur, Illinois. Bob was there enjoying the Christmas Eve holiday with his family. They enjoyed their family gathering with the good foods and singing that are traditional for the holiday, even the lengthy Partridge in a Pear Tree (Twelve Days of Christmas). As recalled by his wife, Dorothy, Bob was in a rare mood that day, having passed along to their granddaughters their heirloom Christmas Village. He was thrilled at how happy they were to receive these gifts (D. James, personal communication, June 2, 2013). Bob found deep joy and contentment in his family and derived much strength from it. It was one of the things he cherished most in his life. On that very cold Christmas Day, as he often did, he went out early in the morning to walk along the trails near his daughter's home. He collapsed on the trail, near the end of his walk, and was later found by a passing couple. Bob was taken to the local hospital's intensive care unit, and 3 days later left his

Going Back for Our Future II, pages 57–68

legacy to us. The crucial point here is that he was doing something that he enjoyed, that invigorated and refreshed him. This is how he lived his life—with vigor and vitality and hope.

Bob pioneered much with respect to individualized science instruction—establishing a significant foundation in it through his early-career study of the Intermediate Science Curriculum Study (ISCS)—coming to understand its role in learner growth and development, and its importance relative to curriculum content. To Bob, the learner was far more important than the curriculum. He applied his depth of understanding through inquiry strategies, which enabled him to significantly impact virtually everyone he encountered. Through his vision, Bob applied approaches and strategies that were to become important to laboratory school curriculum development in the late 1960s and provided critical keys in structuring in-service teacher education and action research paradigms (R. Yager, personal communication, May 7, 2013). Bob became integrally familiar with many content curriculum projects (often called the alphabet curricula), assessment and evaluation tools (such as force-field analysis and the concerns-based modeling), investigatory modules, and inquiry processes. These were among the tools he used with vigor and vitality to help others, injecting hope and self-confidence in them. All this became such a part of Bob that it personified him.

Bob was born on July 30, 1936, in Albany, Missouri, just north and a little east of St. Joseph in northwest Missouri. The family owned a farm there—one that Bob considered returning to after retirement. He attended Northwest Missouri State University where he earned his bachelor's degree in chemistry (1959) and then went on for his masters of science degree at the University of Northern Iowa (1962). His path would later lead him to pursue a doctorate in science education at the University of Iowa, where he was mentored by Bob Yager, completing his studies there in 1969, including the equivalent of a master's degree in geology as part of his PhD work. Interestingly, Bob chose as members of his dissertation committee some statisticians and faculty from science. His dissertation focused on ISCS from Florida State University. While at Iowa, he provided input for what would later become the university's Chautauqua Professional Development program for Pre-K–12 science teachers, and developed a significant "following" of science teachers active with using ISCS (R. Yager, personal communication, May 7, 2013). Bob later utilized his understanding of ISCS and professional development in his work on self-paced instruction with the Salish Project during the 1990s.

The University of Iowa's Chautauqua Professional Development program began in 1986 and continues today. The program was modeled after professional development programs for college faculty that had been sponsored by the American Association for the Advancement of Science and supported by the National Science Foundation during the late 1960s and early 1970s. One outcome targeted by the project was to help teachers become more student-oriented in their science instruction. Included in its structure was individualized, self-paced instruction, recognition of student diversity, classrooms directed by student questions and experiences, and students being active contributors (R. Yager, personal communication, May 7, 2013). Teachers participating in the projects engaged in action research, often assisted by graduate students. This is a large part of what Bob James was involved with through his time with the project.

Bob continued applying his knowledge of the Chautauqua-type program approach during the mid-1970s, when he directed one of the last NSF-supported summer institutes for teachers. That particular institute focused on ISCS teachers, engaging them in learning about the program through doing the project's activities, led by an experienced ISCS teacher. Jim Ellis (personal communication, June 25, 2013), who participated in the 1974 ISCS institute, remembered how Bob enhanced the experience by incorporating content presentations by scientists, through his own presentations on inquiry and research related to the implementation of innovative programs, and by bringing George Dawson (one of the co-developers of ISCS) to meet with the participating teachers to explain the philosophical and psychological underpinnings of the program. During the year following the summer institute, the teachers met and shared their experiences with the program, including presenting their own research projects about it. This same model has been used by numerous science educators over the years, having been found to be highly efficacious. Bob subtly taught it to his students during the times he mentored them through their studies with him. He had them experience it firsthand, knowing this was the most effective teaching technique he could employ.

Bob's teaching experience included 7th-grade science at Iowa City's University Laboratory School (1966–1968), high school science at Odessa, Missousri (1959–1961) and at Cedar Raids, Iowa (1962–1966), and professorship at Kansas State University in Manhattan, Kansas, where he began his career in higher education in 1969. In 1984, Bob accepted a position at Texas A&M University, where part of his responsibilities included directing the Texas Alliance for Science, Technology and Mathematics, a partnership between institutions of education, businesses, and communities; and directed the university's Center of Mathematics and Science Education. Along with his Center work, Bob found it essential to take leadership roles in the NSTA's Building a Presence project. This was a natural outgrowth of the "presence building" he so clearly articulated and encouraged during the early and mid-1980s as he worked with groups such as the Triangle Coalition and industry and business alliances to improve science education. He recognized that the stakeholders were not only the teachers but also the industrial and corporate world, with the ultimate stakeholders being the children. Through his work with the Alliance, Bob was deeply into S-T-S before it really found its footing as its own educational movement. At times, Bob had to operate as the Lone Ranger to push this to the fore. There were always headwinds and resistance to it. Nevertheless, he found ways to offer summer programs for teachers in STEM industries (such as the Texas Teacher Internship Program) to make classrooms accessible to the real world of industries, agencies, and researchers; and to make those entities more accessible for the classroom. He always viewed that as a synergetic relationship, bringing inherent empowerment to both parties. Through his work with the Texas Alliance, Bob secured over $8 million in extramural funding that supported projects such as those in which teachers were immersed in experiential learning, reflective practice, and then returned to their classrooms to apply what they had learned.

Kevin remembers first meeting Bob when entering the teacher education program at Kansas State University in 1972, after spending a year in the paleontology program. Kevin noted,

> In small ways, it seemed almost fateful that Bob would be my undergraduate mentor, as well as my mentor during my graduate work there. Among Bob's joys was woodworking—and I could make sawdust with the best of them. But, unlike me, Bob usually ended up with something more than sawdust to show for his efforts. I guess that it was the process of making the sawdust that helped make some of the bonds between us—just as it was the process of his nurturing of me and others to become science educators—not just science teachers. He clearly modeled what he believed a good science educator and mentor should be and do, and he sought means for us to have opportunities to grow in those same ways. He would not allow us to just make sawdust, but constantly challenged us to look at the potential that lay beyond and examine the processes we lived by and the products that defined who we were becoming as professionals.

At that time, the teacher education program was what was called a "competency-based" program, and students had to demonstrate competence in specified skillsets critical for effective teaching. The process was, at times, grueling, but Bob kept a steady hand in guiding everyone through it. There was no failure with Bob. He would simply not allow it. He was determined to have each student make the best of themselves. As Kevin noted,

> I think he believed that was his duty to us, our duty to ourselves, and our collective duty to our future students and profession. Bob was truly interested in coming to know each of us, what motivated us and guided us in crystallizing our perceptions and understandings about ourselves and our chosen field—and what motivated us as individuals beyond or outside our profession. He

was interested in our lives, in our families, in the holistic view of what made us what we were.

Kevin graduated from the undergraduate program in December 1975 and maintained steady contact with Bob thereafter. In 1978, Kevin decided to begin work on his master's degree. Of course, to whom should he turn but to Bob? As with many students at that stage, Kevin knew he wanted to complete a master's degree, but his focus on a particular thesis topic was not well defined. Bob was rarely without ideas regarding matching his students with projects. One of Bob's attributes was his resourcefulness, and that included his collecting of what seemed as a bottomless pile of science curriculum materials. Not only did Bob go out of his way to collect them, but he actively engaged his students in coming to know them and utilize them. It was through Bob that many students were first introduced to ESS, SCIS, SAPA, the earth science curriculum project ESCP (Kevin's favorite!), ISIS, ISCS, BSCS (all three versions), CHEM-Study, Nuffield Science from the UK, and on almost ad infinitum. Bob actually was serving as a site coordinator for ISCS (Intermediate Science Curriculum Study out of Florida State University) at the time he mentored Kevin and was doing some piloting of the ISIS program in the area as well. The hallmark of these curriculum projects and the common linking thread Bob emphasized was science inquiry. Bob insisted his students learn not only how to learn by inquiry but how to teach with it. Those lessons have held many in good stead throughout their entire professional lives. With Bob's guidance, they built a foundation of sound, inquiry-based science understanding. Inquiry was not just something Bob did to his students. He started them inquiring about inquiry. He didn't just talk about it, but immersed his students in it. Although having students know science content was important to him, Bob was always striving to get them to look beyond the content and see into what was going to really be of benefit to students—dissecting the "how" of sciencing and its importance to the real, everyday lives of his students and the students they would be teaching.

Janice recounts very similar experiences. She had just finished her 4th year teaching in June 1996 and was a participant in one of Texas Alliance's biotechnology workshops. She was excited to be there to learn some new ideas to take back to her classroom, but was more excited to be back at Texas A&M University for 3 weeks and for the stipend that came along with the workshop. Janice visited with Bob during the 3 weeks of the workshop about a lot of different things: her teaching, her personal life, and what she saw herself doing professionally. At that time she hadn't really thought about doing anything other than being a classroom teacher. One day, her chat with Bob was different. As Bob was wont to do, he looked at her with a gleam in his eye and said, "How would you like to come back to Texas A&M

and work on your master's degree with me?" Janice's answer to that question began her path as a science educator—as it did for so many teachers and students—establishing enduring relationships that changed their lives.

For Janice, her graduate work began with Bob giving her the main task of co-teaching his secondary methods course with him. Bob would tell her he needed "reality checks" with someone who could bring recent classroom experience to his methods class since it had been "a really long time" since he had been in a high school classroom. Bob never seemed to gradually move students into their roles. He immediately immersed them. The immersion he started for Janice was to have her assist him with a report. Earlier that year, he had obtained a list of approximately 100 research articles that were considered the "must read" in science education from a session he attended at either ASTE or NARST. Those articles covered everything from the nature of science to pedagogy to technology. Bob gave Janice the list and sent her to the library with a stack of copy cards, telling her to, "Find all of these articles and make a copy for both of us. Then, read each one and write a paper that tells me what the research says to you as a practicing teacher and what you have to say to the research." Bob gave many of his students a similar assignment, always working toward that intersection between research and practice. From those points of intersection, Bob always would seem to return to the question, "How does the research inform practice and how does practice inform research?" Bob strove to get preservice teachers reading science education research and find the value of it in their professional lives. At Texas A&M, those undergraduate students were science majors, so although they were experienced in reading science research, reading science education research was a different thing altogether—and very limited. But, as he did wherever he taught, Bob never relented, and constantly exposed his students to what science education research could offer for the classroom.

Both Janice and Kevin recall how Bob had them model and teach those co-taught methods courses based on "best practices." As good mentors do, Bob would have debriefing sessions in the lab or in his uniquely organized office regularly. (Kevin mused about the depth of the paper piles on Bob's desk and the apparent mess of it—sometimes items extending back multiple years—yet he knew exactly what he had and where on the desk it could be found.) These sessions always examined the success of the modeled practices with the students, examining what were the strengths and weaknesses, and what could have been done differently to make the lesson better.

Bob was well ahead of what educators now call reflective practice. For him, it was always done in detail and at length. No learning session, formal or informal, would pass without time being devoted to reflecting on the experience. Bob believed teacher reflection was critical if transfer of experiences into teaching practice was to actually happen. At times, Bob had to teach his students and teachers (and his friends and colleagues!) how to reflect. He spent time and considerable energies posing questions that prompted them to think more deeply, to connect to prior learning and experiences, to extend their thinking to future applications of practice, and how to ask their own questions about what they hoped to learn, what they had learned, and about what it all meant to their practice. It was "KWL Plus" before KWL became commonplace. His patient guidance in this regard was always present. This meant, of course, sometimes slowing things down in his classroom so that it was possible to set the stage and allow for reflection, allowing for both the time and emotional safety for it to occur. In those days, it was more common to charge ahead and plow through new materials or do more activities to illustrate yet additional ways to teach the same topic, and there seemed to be no shortage of those extra activities. That was not Bob's way. The kind of reflection Bob insisted on having was not always easy. But it became inseparable from instructional and experiential components, becoming embedded in our ways of thinking about what we were doing, why we were doing it, and the consequences (positive and negative) that followed. Reflective practice became a part of us. Of course, this had ramifications beyond science class. Once reflective practice became part of our DNA, it exhibited itself in every other aspect of our lives. In many ways, it put "humane" into "human."

In their conversations with many of Bob's former students, Kevin and Janice found that he always treated them as equals and never made them feel like "lowly graduate students." He valued their opinions, and those opinions really mattered. He nurtured collaborations that would have not been possible otherwise. Those collaborations continued well past graduation, often including invitations to former students to return to his classroom to share their understanding and experiences with his current students.

Bob didn't limit all this to his students, but extended it to his family as well. As his daughter Beth related,

> Science inquiry (and inquiry interaction, in general) was something Bob also modeled within his family as his daughters grew to be adults. He enjoyed conversation, asked good questions, and knew how to draw out his listener—whether speaking with good friends, colleagues, his wife and children, or his grandchildren. All six of his grandchildren lived across the country, and he rarely visited or hosted them without having prepared materials for an inquiry activity in some aspect of science. Of special note were the hot air balloons he constructed with them one summer, an activity that used an ice cream carton to study reflected light, gluing together a three-dimensional star chart, and building a birdhouse. To date, two of Bob's three grandchildren who are high school graduates have pursued science degrees, and Laura, his youngest daughter, followed his footsteps as a secondary science educator. (B. Holmes, personal communication, May 29, 2013)

Although Beth decided to pursue music education as her profession, she has reflected on the ways Bob's emphasis on inquiry has even impacted her approach to teaching music. Her approach is premised on foundational understanding she learned from her father that

> students always learn best by discovery, and taking them into the "underground caverns" of the musical choices a composer makes, to discover for themselves the "jewels" that lie hidden there, with a few tools I supply (questions and a safe environment to pose responses) is a stimulating adventure for all. (B. Holmes, personal communication, May 29, 2013)

Bob had a deep sense of understanding that questions were at the core of successful discovery and inquiry learning. Bob seemed to be a master of asking questions and leading others to answers through questions he would ask and through questions he could get them to ask themselves, so much so that his wife told him that as he would teach, he would draw pictures with his words (D. James, personal communication, June 2, 2013). This questioning was characteristic of Bob in his church and social life too. His friends valued his insights, caring questions, and focused listening.

All of Bob's focus on the importance of inquiry science became even more evident to Kevin when he entered the doctoral program with Bob as his advisor. Kevin reflected,

> It was during this time that I really began to see the breadth and depth of Bob's commitment to inquiry instruction and became aware of the tools he brought to the fore to help facilitate it. Looking back on that time, it is evident to me now that Bob knew exactly how far to push his students, how much they could absorb and cognitively reconcile, and when the best times would be to do that pushing. He could have exposed us to these inquiry tools in previous years during our professional formation, but we would not have understood nearly as well the underlying structures and utility of them. Bob had a keen sense of when to immerse us ever more deeply into it.

One of those tools was a combination of learning process, learning information, and experiencing firsthand how those things functioned when working with teachers. For Kevin, this was the first graduate course for teachers in which he was an assistant. In keeping with Bob's intense focus on inquiry instruction, he and Larry Enochs structured the course to engage the teachers in inquiry learning applicable to both in-classroom and out-of-classroom contexts. Bob gave Kevin several charges. One was that of taking notes, and from those notes generating summary documents the teachers would find useful. A field-trip planning document was one example. Another charge Kevin was given was to research, plan, and deliver discrepant events and demonstrations. Kevin remembered, "This was my first foray into discrepant events, and Bob knew when I was ready for it. Together, we learned how these short engagements fueled inquiry learning." Similarly, Janice's journey with inquiry was guided by Bob. She recalled, for example, an activity she did with him called Gobstoppers. As for many of his former students, whenever Janice does one of those activities, she can still see Bob's grin and the sparkle in his eyes, and can hear him say, "Well, why do you think this happened?"

Related to the teacher graduate course, Bob introduced Kevin to many "current" developments in research ideas and instrumentation, and one of those was what was called "force-field analysis" and the other was the Concerns-Based Adoption Model (CBAM) out of The University of Texas-Austin. Bob understood that tools like the force-field analysis and CBAM's Stages of Concern would provide useful information that would guide him in how to continue to structure the graduate course to make it as effective as possible for the teachers. Bob spent some time at The University of Texas working with Gene Hall, Archie George, William Rutherford, and Susan Loucks (later Loucks-Horsley) learning about CBAM and how to utilize it. He returned to Kansas State and actually implemented a course to teach doctoral candidates in its use, from the Stages of Concern, Levels of Use and Innovation Configurations. Again, Bob's timing was impeccable, since he had a clear sense of when his students would be ready to learn about these things. Much of it was, by Bob's direction, planning, and foresight, contextualized in getting teachers (and students) moving forward with inquiry-based science.

Looking back at CBAM, one of the significant pieces relative to learning to use it was developing usable rubrics that had application for Levels of Use and especially for Innovation Configurations. Users would identify key elements relative to an innovation, and then, based on research and best practices, write descriptors that illuminated differing levels of quality of use of each element along a continuum from ideal to acceptable to unacceptable. As familiar as these structures sound today, they were novel and groundbreaking in their time. Bob made sure his graduate students learned how to use these kinds of rubrics and how to construct some themselves. He utilized those kinds of rubrics for his own courses and had his students incorporate them too. This knowledge and practice positioned his students well when national standards emerged and were promulgated along with their attendant assessments.

Bob also created an undergraduate science education course entitled, "Laboratory Techniques for Science Teachers." That course was designed for all secondary science education majors. He developed modules for the course that focused on how to perform specific techniques in both field and laboratory settings and teach students how to do them. In all, there were over 30 such modules he generated. From this work, Bob, along with

Frank Crawley, who was at The University of Texas at Austin at the time, published a manual for science teachers to prepare them with the skills and techniques to be effective laboratory science teachers (J. Ellis, personal communication, June 25, 2013). There were, of course, standard laboratory exercises in the sciences, some of which were extended from those identified as essential for students to complete for entry to college in the late 1800s. As the field of science education began establishing itself, students essentially took their science and science laboratory courses and then some pedagogy courses (mainly clinical experiences) to complete their training. To Bob, this approach contrasted starkly with what he saw as the need for teachers to be educated about the use of the laboratory as an essential and central pedagogical strategy and come to understand appropriate applications of it for a variety of learning needs that made more sense cognitively for children. However, Bob recognized most teachers would be unlikely to create laboratory investigations meeting those criteria, but would be likely to make use of some if they were provided. So Bob undertook the task of developing science laboratory manuals, many of which could be used in individualized instructional settings that would fit with his vision. In a real sense, Bob was pioneering the science education laboratory at the same time similar efforts were underway with what we refer to as the alphabet curricula. As he developed and published the manuals and modules, Bob incorporated them into his own courses. Bob believed that students should have this knowledge in their arsenals to effectively approach inquiry learning. He insured his graduate students knew the investigative and laboratory techniques by having them engage in his courses in some way—either by completing modules (and helping develop more) or by actually supervising the course. Again, his specific aim was to enhance student and teacher abilities with doing inquiry in appropriate ways.

Wherever Bob was, he made it a point to be active within the profession and at multiple levels (state, regional, and national). His goal was to inculcate within the culture of science education the most effective (and what we now refer to as "best practices" of) science education. He regularly presented at NSTA and at southwest ASTE meetings and sought opportunities for his graduate students to be engaged there as well (in fact, he expected them to be engaged and attend such meetings, stating that, "That is how you meet people and get your name out."). Bob found it refreshing to go to such meetings and present, since it gave him yet another way to stay in touch with former students and colleagues at other institutions (D. Janke, personal communication, June 17, 2013). He always seemed to find time to have coffee or lunch or dinner with colleagues and former students. He could effortlessly draw others into conversation that intertwined professional and personal life in the seamless manner in which life actually occurs. He was always interested in what others were doing and what was going on—and how it was affecting them. He applied this same strategy at the state and local levels, building networks of science teachers, and through his work through a variety of professional organizations (including AETS/ASTE, NSTA, the School Science and Mathematics Association, the Triangle Coalition, etc.). Whoever Bob was meeting or "just visiting" with, he always made it seem as if that person was of the utmost importance and showed keen interest and concern in him/her, both on a professional and personal level.

Bob consistently reached out to colleagues at any level within the university and across colleges as well. While at Texas A&M University, for example, Bob teamed with Delmar Janke (whose major focus was on elementary school science) and later Carol Stuessy (whose focus was middle school–level science) to bring ever-increasing focus on inquiry instruction to teachers in the field (D. Janke, personal communication, June 17, 2013). Bob's influence was evident through multiple invitations he consistently received to share and present at local schools throughout the state. He was a fixture at meetings of groups such as the Texas Science Education Leadership Association, paralleling his earlier work when at Kansas State University and his involvement with the Kansas Association of Science Teachers.

Bob's contributions were recognized at several points in his career. He was known as a leader in science teacher education among his peers and colleagues. For example, he was honored at Texas A&M University by his former students with the Texas A&M Distinguished Achievement Award in 1999, right on the heels of being awarded the Texas A&M Extraordinary Service Award by the College of Education Development Council in 1998. Jim Ellis recounted how Bob was one of a core group of science teacher educators who was a leader in establishing and supporting the Southwest Region of the then-Association for the Education of Teachers of Science (SWAETS). Jim said,

> I was introduced to the group at a meeting in Estes Park, Colorado, in January, 1971. It was an informal group, and as a graduate student, I was told by Bob and others not to miss the midnight seminar that was on the agenda. At some point in the evening, being a serious graduate student, I asked when the seminar was to begin. That was the point at which I learned the most valuable opportunity at any conference is the time spent visiting informally with colleagues. Bob James is among these colleagues who have contributed the most to my development as a science teacher educator. (personal communication, June 25, 2013)

Those midnight seminars are still talked about at ASTE conferences. Bob also served as the president of the Association for the Education of Teachers of Science (AETS, now ASTE) in 1986–1987 and was honored with that organization's Outstanding Mentor Award in 2000. These are among the obvious recognitions one might find in perusing his curriculum vita. What is hinted at through the ASTE Outstanding Mentor Award is the unwritten recognition he received throughout his career from his students (many of whom maintained close friendships with him once they graduated) and colleagues with whom he worked daily. We're sure Bob found that the most satisfying recognition of all.

Bob was interested in life, and vigorously pursued that interest. He coupled his love of science with his love of the arts, particularly music. Perhaps that, in part, led to his daughters' love of music and their choices to pursue it as their vocations. Bob married Dorothy Evans of Savannah, Missouri, in 1959, with whom he had three daughters: Beth, Carolyn, and Laura. From our occasional discussions with him when we'd catch up with each other at professional conferences, we know they, and his six grandchildren, were songs in his heart. Bob was a man of deep faith—in himself, in others, and in a religious sense—striving to help each of us be better persons and to come to understand ourselves much more fully than we would have thought possible. That was the driving motivation for Bob. That was what life was all about. That is what will carry forward from Bob for many years to come.

ACKNOWLEDGMENTS

We gratefully thank the following individuals for the time, energies, and information they provided that enabled us to write this chapter: Dorothy James, Beth (James) Holmes, Jim Ellis, Delmar Janke, Robert Yager, Caroline (Beller) Fye, Tim Scott, and Adrienne Bentz.

CHAPTER 4

JANE BUTLER KAHLE

Her Journey of Transformations and Career Curves

Dana Riley Black
Institute for Systems Biology, Seattle, Washington

G. Nathan Carnes
The University of South Carolina

Malcolm Butler
University of Central Florida

Kathryn Scantlebury
University of Delaware

The authors of this chapter—three of Jane's former PhD students and one scholar of her work—were compelled to compose and share the story of Jane's career because of the ways in which she impacted their lives and careers, as well as countless science educators, teachers, and students nationally and internationally. In addition to being drawn to Jane because of

Going Back for Our Future II, pages 69–86

her efforts to address inequities in science education for students, teachers, and researchers, the authors recognized and have great respect for her intellectual ability, research acumen and experience, and her reputation as a supportive and rigorous mentor.

Indeed, Jane is an acclaimed *leader* in science education, as shown by the depth of her research, scholarly work, and accomplishments. Since the early 1970s she has served as the Principal Investigator for 80 research and training projects (monies over $56 million), Co-Principal Investigator for three, from 18 different funding sources. Jane has published over 170 articles in peer-reviewed journals, 48 chapters in books or monographs, eight editorials, six books, and three Congressional testimonies, as well as numerous research abstracts, technical reports, and position statements. Jane is the recipient of over 27 honors and awards, including Purdue University's Distinguished Alumni and Outstanding Woman Scholar awards; Miami University's Distinguished Scholar award; the Benjamin Harrison Medallion; an Honorary Doctorate of Humane Letters; the American Education Research Association's Willystine Goodsell award; the Distinguished Contributions to Science Education Through Research award from the National Association for Research in Science Teaching; and the Distinguished Service to Science Education Citation from the National Science Teachers Association. Further, she was elected a Fellow of the American Association for the Advancement of Science and a National Associate of the National Academies of Science and Engineering.

This chapter will attest to Jane as not only a leader in science education but also a *pioneer* in science education. As will be addressed, throughout her career, Jane was one who embraced opportunity to *venture into new territory*, or areas of study and practice in science education (e.g., post-Sputnik Teacher Training Institutes, serving as Purdue University's first female science/math education tenure-track professor, leading the development of one of the nation's first Statewide Systemic Initiatives, and serving as the first chairperson of the National Research Council's Committee on Science Education). As Jane ventured into new territory, she *fostered and advocated for new areas of research and practice* (e.g., equitable teaching practices for girls and African American students, research methods to expose and explore inequitable learning opportunities, and the application of teaching practices and research practices to inform education policy). And importantly, Jane *prepared others to follow* her in these new areas of research and practice, serving as the major professor for 25 doctoral students and as a committee member or reviewer for 50 others. And undoubtedly, she has impacted many beyond the students she directly mentored.

To frame the chapter, we start with a quote by Norton Simon (1907–1993) an industrialist and philanthropist who based his work in California. Simon was a close friend of Michael E. Phelps, the inventor of the technology behind the PET scan. Before his passing in 1993, Simon provided Phelps with a

> precious present in a simple statement... "Life has a natural curve. You go up, plateau, and you go down. The only way to change this is to be continually starting new curves, and in this way, always stay in the state of becoming." (Phelps, 2004, p. xiii)

Simon's quote increasingly is cited by leading scientists (e.g., Leroy Hood and James Heath) to describe the historical trajectory of their own careers, and as inspiration to guide those whom they mentor.

Most certainly Simon's quote can be used to describe Jane's life and career as a pioneer in science education. Jane's career did not follow a traditional curve, a natural curve, or a bell curve, as it did not plateau; rather, it continued on an upward trajectory with several points of *transformation* that produced a series of new *career curves.* As will be described, Jane's transformation points were characterized by her ability to address challenges directly, leverage opportunities, and foster strong collegiality through her networking, relationship building, and willingness to utilize advice from others. And her career curves were characterized by strong vision and commitment, interest in the continual update of research skills, as well as multiple modes of dissemination of research results and outcomes.

Career Curve: Classroom Teacher (1956–1969)

Jane graduated from Wellesley College with honors in botany and zoology. While she studied to be a scientist, Jane enjoyed tutoring at a time when most women could not envision themselves as graduate students in science. Subsequently Jane switched her studies to science education, completing high school biology certification requirements at Teachers College, Columbia University (New York City) and Purdue University (Lafayette, Indiana).

As Jane began her career as a teacher, the district superintendent informed her that she was too young to teach high school students (specifically, to teach boys). As a result, Jane began her teaching career at Washington Elementary School on the banks of the Wabash River (in Indiana) teaching 4th- through 6th-grade science and health. The following year she replaced her former biology teacher and for 10 years taught biology in rural high schools. Already contemporary in her practice, Jane demanded to use the new BSCS textbooks with her students, while preparing herself by memorizing the BSCS *Handbook for Biology Teachers* (BSCS, 1963). Jane described her time as a classroom teacher as "a wonderful ten years, interspersed with two maternity leaves." (Kahle, 2007).

Transformation: Sputnik

The launch of Sputnik in 1957 ignited a "crisis" in U.S. science education, resulting in the National Defense Education Act, which included the National Science Foundation's teacher training institutes. To advance her career, Jane applied for and was accepted into a biology teacher training institute at Purdue University (1961–1963), however she did not receive financial support because she was married and had two children, which fostered the assumption by the director of the program that she would leave the teaching profession.

Career Curve: Graduate Student (1970–1972)

Opportunity to continue her own educational advancement did not end at her completion of a Master in Science degree at Purdue University. While at Purdue University, she met her first true mentor and master's advisor Joe Novak. Joe moved to Cornell University and encouraged Jane to apply for a Shell Merit Fellowship at Cornell University, which she attended for two summers, first as a participant (1968) and second as an instructor (1969). During her summers at Cornell University, Joe introduced Jane to key stakeholders in science education and encouraged her to continue expanding her network and studies.

Such networking brought Jane to Sam Poslethwaite, a well-known botanist in the Department of Biological Sciences at Purdue University. Sam became Jane's doctoral advisor. Sam had been very successful in teaching large sections of botany to undergraduate students through an individualized approach, called the Audio Tutorial Laboratory (ATL) (Belland, 1991). ATL placed audiotapes, Super 8 millimeter projectors, and live materials in a study carrel in which students would progress through materials at their own speed (i.e., asynchronous time).

In the early 1970s, the science departments at Purdue University fostered an exchange program among their faculty and outstanding teachers in Gary, Indiana. Floyd Nordland, one of the exchange faculty was interested in implementing ATL with his school district partners at Gary West Side High School. Both Jane and Floyd were interested in adapting ATL for high school biology classes, especially in schools enrolling underachieving African American students. Jane was particularly keen to understand how adequate time to learn would affect any achievement gaps between groups of students. Moreover, having grown up in a small town in Indiana with an awareness of and a tenacity to "buck" prejudice, she was particularly interested in the possibility that ATL would provide underachieving African American students an opportunity to learn science.

Jane and Floyd adapted the ATL materials to portable carrels built by a colleague and went first to Gary, Indiana, and later to Chicago, Illinois, to test the ATL in high school biology classrooms (Kahle & Nordland with Charles & Sims, 1974). The success of the project and the research was the foundational inspiration for Jane to continue pursuing a career focused in the area of equity and science education.

Transformation: Government Review of Research Universities

Directly after completing her doctorate, Jane was employed by Purdue University as an instructor of biological teaching methods and to supervise biology student teachers. Soon after Jane began, the federal government conducted reviews to identify evidence of discriminatory practices affecting women and minorities at major research universities. If such evidence were found, federal research funds would be withheld from the institution. Without any discussion or notice, Jane received a letter from Purdue University retroactively changing her appointment from a non–tenure-track instructor to a tenure-track assistant professor.

A complementary point of transformation was advice Jane received from Bob Kane during an annual performance review. Bob, a mathematics educator and chair of the Education Department at Purdue University, suggested that she think about where she wanted to be in 5 years and that she prepare a plan to achieve this goal. He also suggested that she select a focus for her research and that, if she stayed with that focus, she eventually could become a national and international expert in the field.

Career Curve: Professorship and Focus on Gender Equity (1972–1989)

(1972–1989, Professor; 1984–1986, Professor and Associate Dean of the Graduate School)

Following Bob Kane's advice, Jane decided to focus her research on equity to include the emerging field of gender equity, specifically girls in science and mathematics. Jane decided to focus on gender equity because issues of inequality had impacted her own career. Similar to other science/math educators at Purdue University, Jane had a joint appointment in a disciplinary department as well as the education department, with her disciplinary appointment being in the biological sciences. The difference between Jane and the other eight science and math education faculty was that she was the only female, and she fought for equal treatment.

During her tenure at Purdue University, Jane advanced to become one of the leading national and international researchers on equity issues and science education. Her initial research on students' attitudes toward science focused on African American students. In the 1980s, Jane identified that this racial group of students studying in low-income schools had more positive attitudes toward science, but fewer opportunities than White students had to engage with science, both in and out of school. Jane, along with women professors at historical Black colleges and universities (HBCUs) who were participants in Jane's project to enrich the professors' research experiences, highlighted this issue in the report, *The Double Dilemma: Minorities and Women in Science Education* (Kahle, 1982); that is, the challenge that students from underrepresented groups faced if they wished to participate in science or science education, especially if they were female and African American.

In the early 1980s, Jane and her doctoral student Marsha Lakes (now Matyas) used an equity perspective to examine the data for science education from the National Association of Educational Progress (NAEP) concerning students' opportunities to learn. Their article, "The Myth of Equality in Science Classrooms," showed that 9-year-old girls, compared with boys, reported fewer in-and out-of-school science experiences (Kahle & Lakes, 1983). Moreover, classroom experiences did not close the gap. At that time, girls' lack of science experiences was attributed to their "lack of interest," "low risk-taking behaviors," and such. However, Jane and Marsha's analysis of what girls would like to do indicated that girls wanted to do more science activities both in and out of school. The article was one of 13 papers selected for the 40-year anniversary issue of the *Journal of Research in Science Teaching* and has been one of the most frequently cited papers in science education (Holliday, 2003). This seminal work spearheaded Jane's research examining girls' science classroom experiences.

The "myth" analysis prompted Jane to engage in collaborative, case study research focused on identifying teachers' practices and strategies that encouraged girls in science. This research showed that science teachers in Indiana, Florida, Colorado, Louisiana, and other states, who were successful in encouraging girls, found ways to connect science to students' interests (Kahle, 1983). Jane found that other common practices of these teachers included integrating laboratory work into the curriculum; showing students the connection between science concepts and laboratory work; use of frequent, regular assessments to monitor students' progress; and proactively serving as a career advisor to students by way of explaining why advanced mathematics and science courses were important and intervening when students, especially girls, chose a pathway that could prevent their achievements in science. This research informed the writing of *Girls in School: Women in Science* (Kahle, 1983), highlighting how individuals (teachers) could challenge and change the gender structure within science.

Jane's critical review of students', especially girls', opportunities to learn science, led her to Mary Budd Rowe's research on wait time (Rowe, 1986), which then connected her with Australian science educators, Kenneth Tobin, Lesley Parker, Leonie Rennie, and Barry Fraser. Jane's research was soon recognized at the international level as she collaborated with Lesley and Leonie to conduct gender studies in Australia and the United States, and as she joined Ken and Barry to lead a research group studying science teaching in Australian secondary schools. Their findings were published in *Windows into Science Classrooms: Problems Associated with Higher-Level Cognitive Learning* (Tobin, Kahle, & Fraser, 1990). Through this research, Jane and her colleagues showed that the equitable science teaching practices that encouraged American girls in science worked in Australia. They also discovered, however, that teachers were often unaware of their gender biases as they used subtle, sexist practices to ostracize girls from science.

The abovementioned work—*Double Dilemma* (1982); *The Myth of Equality in Science Classrooms* (1983); *Girls in School: Women in Science* (1983); *Windows Into Science Classrooms* (1990)—led Jane and colleagues to develop a model to test whether gender differences in science education were due to girls' characteristics or their response to teaching/learning environments. Jane, Lesley, Leonie, and graduate student Dana Riley (now Riley Black) proposed a model that interwove sociocultural influences with those found in classrooms. That paper, *Gender Differences in Science Education: Building a Model* (Kahle, Parker, Rennie, & Riley, 1993), provided an impetus for gender research in science education similar to the catalytic effect of Eccles and Fennema's models in mathematics education (Eccles, 1983, 1989; Fennema, 1981; Fennema & Sherman, 1978). As with her earlier studies of African American students, the development of the equity model and the comprehensive review of the research literature that showed teachers' instruction, expectations, beliefs, and morale were critical indicators in narrowing achievement gaps and retaining students in science.

Without a doubt, this period of Jane's career established her as a pioneer in science education, leading the field into new areas of research on gender equity. Additionally, Jane worked to prepare others to apply this new learning into practice and to extend the research agenda. For example, as she supervised student teachers, she continued to combat the belief that African American students could not learn science as well as White students. Jane encouraged student teachers to leave their "safety zone," proposing the opportunity to accept placement in schools with high African American populations such as many in Gary and Indianapolis, Indiana.

Jane's mentoring of graduate students has had a lifelong impact, as described in a testimonial by one of Jane's Purdue University doctoral students Kathryn (Kate) Scantlebury, who now serves as professor of chemistry and biochemistry at the University of Delaware.

> When I was a high school chemistry teacher in Perth, Australia interested in pursuing a PhD, one of my professors, David Treagust advised me to make a list of researchers I would like to work with and a list of excellent science education programs—in an ideal world there would be over lap between the two lists. At the top of my list was Jane Butler Kahle. Jane was visiting Perth, and I met her at the annual Western Australian Science Teachers' conference in 1986. An hour into our first conversation, Jane provided me with the "opportunity of a lifetime"; that is, an invitation to immigrate to the US and study with her.
>
> A phrase I associate with Jane is "niceness counts!" The stereotype of an academic can be a person (usually a white male) extracting high quality research from the group through helpful, albeit, negative critiques and abrasive reviews. A research group operating in a hierarchical structure with clearly defined roles from the professor down to undergraduate researchers. Jane's focus on gender roles and stereotypes always provided an opening for strong reviews and critique of one's work, but in an environment that also embraced caring for one as a scholar and a person. She taught me to embrace feminist research and practices, while remaining cognizant that excellent scholarship emerged from a theoretical foundation; the selection, collection and analysis of appropriate data; the clear, articulate dissemination of outcomes through a variety of avenues and outlets; and respect for all participants—students, teachers, peers, staff and oneself. (K. Scantlebury, personal communication, 2014)

In addition to the formal academic mentoring pathways, Jane has worked with numerous science teachers, school district personnel, and faculty. Jane is a proactive mentor and constantly seeks ways to ensure her students, the teachers she collaborated with, and colleagues had opportunities to engage in science education through continuing their education, developing skills and knowledge, and establishing and extending networks. Moreover, as one of the few senior women in science education, she was an informal mentor to many scholars in science education, especially women, who sought advice on the challenges of melding professional careers and personal lives. Malcolm Butler, associate professor at University of Central Florida, describes an example of Jane's influence in a testimonial.

> My initial knowledge of Jane and her work came through an interaction with Dr. Mary Budd Rowe, during my graduate school days at the University of Florida in Gainesville, in the early 1990s. Finding out about wait time led me to Jane's scholarship on equity, especially as it relates to Black children. Several years later, I ran across the Equity Metric [Kahle, 1998a] that she helped develop during her time at the University of Wisconsin.
>
> Jane and I finally met in person when she was a plenary speaker at a symposium for science education scholars of color that was sponsored by a NSF grant for which I was serving as Co-Principal Investigator. Jane and I talked at length a couple times over the course of the event, and I observed numerous

> interactions between her and the symposium participants. It was then that I experienced her calm demeanor and passionate personality, which she is able to combine so well.
>
> Jane was so well received at the first symposium that she was invited to make a return engagement, which she did in 2013. I continue to use her as a role model for how to be a citizen scholar and champion for the marginalized in mathematics and science education. (M. Butler, personal communication, 2014)

In addition to serving as a mentor, Jane has always acknowledged the significance that mentors played in her career—in part her dedication to mentoring is "payback" to those who mentored her. In addition to Postlethwaite and Novak, mentioned earlier, she credits Mary Budd Rowe for showing her that a woman could be a first-class science educator. She also acknowledges the role that Paul DeHart Hurd played in asking her to join Rodger Bybee and Bob Yager as part of a team studying biology education nationally—what Jane calls "my first big time consulting" (J. Kahle, personal communication, 2014).

Transformation: Move to Ohio

Jane's success at Purdue University was noticed and in the 1980s she was sought as a candidate for the position of Dean of Education by six universities. Administrators and faculty at Miami University actively recruited Jane to its newly endowed professorship in science education and, in 1989, she accepted that position in Miami University's School of Education and Allied Professions (now the College of Education, Health and Society). This move provided her with a significant salary increase, extensive space for research and graduate students, clerical support, and financial support for the two PhD students who followed her but who were completing their degrees at Purdue University, as well as the political environment to address large-scale change in science teaching and learning.

Career Curve: Systemic Reform and Influence on Education Policy (1989–2008)

(1989–2006, Endowed Chair; retired in 2006 and returned; retired again in 2008)

Soon after her move to Ohio, Jane collaborated with Kenneth Wilson of The Ohio State University (OSU). Together, Jane and Ken wrote a Statewide Systemic Initiative (SSI) proposal to the National Science Foundation (NSF), which was supported by both state educational agencies and the governor. Ken, a Nobel Laureate in physics, brought a great deal of legitimacy to the proposed project, as he was well known for his work in theoretical physics and had a growing interest in science education, while Jane contributed the foundation of leadership in the field of science education, along with the personal and professional skills to enlist the involvement of a wide range of stakeholders.

In 1991, when the SSI entitled, *Project Discovery*, was awarded, Jane's research moved to a macrolevel, as she assumed the responsibility as the project's Principal Investigator. *Discovery* was one of the first 10 SSI projects funded by NSF, and the only SSI focused on equity in terms of gender, race, and socioeconomic levels. *Discovery* would include ongoing evaluation as well as research-based impact studies, providing Jane the opportunity to conduct some of the first gender *and* race analyses of students' attitudes toward achievement levels in science. Some 20 years after she conducted her research on how good science teachers provided all students opportunities to learn, Jane showed that "standards-based teaching practices" (referred to as "inquiry-centered" instruction in *Discovery*) positively impacted urban, African American, middle school students' science attitudes and achievement (Kahle, Meece, & Scantlebury, 2000)

Leadership of *Discovery* would be Jane's first sustained effort to work in the area of policy. Systemic reform in Ohio provided a daunting challenge in that there are regions of the state that are very rural and regions that are extremely urban (Kahle & Woodruff, 2011), as well as many suburban communities. To add to the complexity, Ohio's education system had multiple forms and layers of governance and structures that made broad-based reform challenging. For example, it was not uncommon for parochial schools and exempted village, city, and county school districts to co-exist within the same geographic area and share the same resources, often placing a strain on support needed for necessary change and high-quality instruction.

The *Discovery* project focused on middle school science and mathematics teachers, based on the rationale that all students were still taking math and science at the middle grade levels, and it centered on economically disadvantaged urban and rural schools and Ohio's largest minority group, African American students. Priority for participation was given to teachers from the 41 school districts identified by the Ohio Department of Education as "at risk." In addition, the project established and supported 12 regional districts, each led by a director and a leadership team. Each team was composed of one representative from industry or higher education and two teacher leaders. Teacher participants were selected by the teams to attended professional development institutes in their regions. The structure of *Discovery* placed teacher leaders in unprecedented roles. For example, these educators had opportunities to work collaboratively with biologists and physicists to plan professional development experiences for classroom teachers.

Jane, the project team, and faculty at Miami University and OSU used research on professional development to design the teacher learning component of the SSI. The professional development consisted of 6-week, content-based summer institutes that employed and modeled inquiry-centered and cooperative learning pedagogy. The summer institutes were followed by four to six professional development sessions during the academic year as well as classroom visits made by regional leadership teams. The academic year activities stressed authentic assessments and equity issues. In the SSI's third year, an equity team was identified to provide guidance and expertise for the professional development instructors and teachers.

During the initial summer, the institutes focused on physical science, and the project adopted the University of Washington's very successful curriculum, *Physics by Inquiry* (McDermott & Physics Education Group, 1995). Physics faculty at Miami University and OSU were supported by the project to attend sessions at the University of Washington for an in-depth experience with inquiry. Teams of faculty across the state were recruited to adapt the Physics by Inquiry approach in developing comparable institutes in biology and mathematics. In *Discovery's* second year, teacher participants could attend one of three 6-week summer institutes, Mathematics by Inquiry, Life Science by Inquiry, or Physics by Inquiry. Members of the leadership teams also joined the institutes and learned side by side with the teacher participants. The professional development instructional teams were composed of university science and mathematics faculty paired with education faculty. After the initial summer, all three institutes were offered at all 26 public as well as over a dozen private colleges and universities across Ohio.

During the six, one-day academic year seminars, the teachers focused on issues associated with inquiry instruction, including gender and ethnic equity, alternative and authentic assessment, uses of technology, and the implementation of inquiry teaching strategies. The general sequence of activities included presentations from expert educators, discussion of issues associated with implementing inquiry instruction, and presentations of sample lessons used by the teachers. Throughout the academic year, teachers developed portfolios that documented changes in their teaching practices. These collections of evidence were helpful in developing rich vignettes that corroborated the findings of observations, questionnaires, and other methods of evaluation.

Not only did Jane lead the design and implementation of the programmatic aspects of the SSI, she also led the initiative's research efforts to ascertain the project's impact and effectiveness, providing multiple indicators to document the success of the SSI. External evaluations of *Discovery* (HRI, 1994, 1995) as well as the project's studies indicated that teacher participants changed their teaching by incorporating the materials and instructional strategies introduced in the institutes. Notably, research by two of

Jane's graduate students, Nathan Carnes and Arta Damnjanovic (Carnes & Damnjanovic, 1993), external evaluation reports (HRI, 1995, 1996), and an assessment of mathematics and science education reform in Ohio (Kahle & Rogg, 1997) suggested that teachers implemented inquiry-centered (standards-based) teaching and learning practices in their classrooms. Jane and the SSI team found that the professional development activities transferred to teachers' classroom practices, and this research supported a change in view about the type of professional development that impacts practice. Principals of *Discovery* teachers responded that long-term, content-intensive, professional development was more effective in changing teacher practice and in raising student achievement than nine other common professional activities, including professional development during the school day (Kahle & Rogg, 1996).

The evaluation and research-based outcomes of the SSI were published in refereed journals to inform the education research community. And with Jane's leadership, the outcomes were formatted and distributed in a manner to inform and influence policymakers. Most noteworthy were the *Pocket Panoramas* (e.g., see Kahle & Rogg, 1996), which were developed and distributed from 1995 to 1999 to all school district superintendents (614), all members of Ohio's General Assembly, all of Ohio's Congressional Representatives and Senators, the governor, the Superintendent of Instruction, the Chancellor of the Board of Regents, key Ohio Board of Regents, and Ohio Department of Education staff, as well as university lobbyists across the state. These small booklets, designed to fit into a pocket, painted the landscape of science and mathematics education in Ohio by presenting data about the scope of *Discovery* (numerical and geographical), changes in instruction, changes in achievement, and changes in attitudes in simple graphic form. Although the data were complex and involved a carefully selected group of matching teachers, the message was clear and succinct. This outreach was a key reason *Discovery* became a line item in the state's budget, extending the impact of the SSI well beyond the 5 years of NSF funding. Although the project has changed focus over the years, it continues to be funded by the State of Ohio—a life span of 24 years. Under the leadership of Terry McCollum, the project used the Internet to establish professional development and a supportive statewide network of science and mathematics teachers. At the end of *Discovery*, Jane was awarded a large NSF research grant to study the results of statewide systemic reform. This statewide project, *Bridging the Gap: Equity is Systemic Reform*, developed a three-part research plan to document the outcome of sustained, content-based teacher professional development.

As was the case when Jane was faculty at Purdue University, while at Miami University she continued as a pioneer in advancing the fields of equity and education through influencing policy. At the same time, she continued to mentor and prepare others for their work in the field of science education practice, research, and policy. The following testimonials by two of Jane's PhD students, Nathan Carnes, associate professor at the University of South Carolina, and Dana Riley Black, Director of the Logan Center for Education at the Institute for Systems Biology in Seattle, depict the foundations and growth of Jane's pioneering work.

Nathan Carnes:

During the 1991–1992 school year, I was in my 12th year of teaching junior high physical science. My former undergraduate advisor at Miami University, Johnny Hill, invited me to participate in a focus group at Miami University with the purpose of informing the launch of the SSI. Shortly afterwards, Jane called me and extended an invitation to join the project team and enter a PhD program at Miami University. Although I was apprehensive, after only a few conversations, she convinced me to embark on this new endeavor. Jane's professional activities and successes were fueled by purpose and passion. I was inspired by those characteristics, while learning many important technical skills (i.e., course proposal development, grant writing skills, working with teacher candidates and professionals) and gaining knowledge of equity and diversity issues related to science education that now serve me well in the professoriate. Without a doubt, my professional life would have taken a different trajectory had she not insisted on me joining Ohio's SSI in her gentle persistent manner. She pushed me beyond my comfort zone to grow in a manner that has allowed me to contribute to science education in more ways than I could ever imagine.

During my tenure with her, I found her ability to persuade individuals to join her in the campaign to elevate the quality of science and mathematics education throughout Ohio was one of her strengths. I am persuaded that it is the reason why so many classroom teachers, school administrators, university faculty, mathematics and science professionals from industry, and legislators joined her to bring the vision of Project Discovery to fruition. (G. Nathan Carnes, personal communication, 2014)

Dana Riley Black:

My father provided me with two pieces of important advice that ultimately provided my introduction to Jane. After my undergraduate study, my father advised me not to go directly to graduate school, but to get some real world experience to inform my choice of further study. As a result, I took a job as a Science Education Associate at the Pacific Science Center in Seattle, and through the Center's outreach program had the opportunity to teach science all across the state of Washington. During one trip across the state, an elementary boy confronted me claiming I could not teach him science because I was a girl, and, girls could not do science.

> Soon I recognized I wanted to pursue graduate work related to girls learning science. A second key piece of advice from my father was to identify a mentor for graduate school, as that mentor would be a very important party of my life. He could not have been more correct. As I began reading journal articles I learned of Jane's work, and coincidentally she was the keynote speaker at the Washington Science Teacher Association annual conference. Soon after meeting Jane, I moved across the country to study with Jane as she was just launching Project Discovery.
>
> Working with Jane and Ohio's SSI taught me the perspective of attending to the "whole school system" as a core strategy for attending to equity. As I learned this strategy is not purposed solely for addressing gender equity, but to address equity as related to many types of labels and variables of our school systems. Importantly, Jane not only taught me the essence of scholarly work, but also the significance of relationships and networks of relationships that are manifested across a school system. This "systems approach" has guided my personal and professional life. Perhaps what I value most about Jane was provided in foresight by my father, in that today, decades after formally graduating, I can continue to call upon Jane as my mentor, advisor and friend. (D. Riley Black, personal communication, 2014)

Transformation: End of Systemic Initiative

Jane grew increasingly aware of the importance of policy in educational reform as *Discovery* progressed. In addition, she became acutely aware that current doctoral programs in science and mathematics education provided neither guidance nor experiences in dealing with local, state, or national policies. She became determined to try to remedy this deficiency in the next stage of her career by service on national committees; by continuing to publish in ways accessible to science educators, teachers, *and* policymakers; and by influencing policy at the National Science Foundation.

Career Curve: National Authority in Science Education

As Jane completed her role as Principal Investigator for the SSI, *Discovery*, she had undoubtedly become recognized as a leader and pioneer in science education. This status affords a scholar of her stature the opportunities to reflect and knit together expertise garnered through his/her career as well as to broadly apply his/her expertise as a national authority. In 1996–1997 Jane accepted an invitation to join the National Institute for Science Education (NISE), University of Wisconsin, as a visiting scholar. And, from 1999 to 2002, she served as Director of the Division for Elementary, Secondary, and Informal Education at the National Science Foundation.

In 1994, Dorothy Gabel, editor of the National Science Teachers Association's *Handbook of Research on Science Teaching and Learning*, asked Jane and colleague Judith Meece to contribute a chapter entitled, *Research on Gender Issues in the Classroom* (Kahle & Meece, 1994). Jane and Judith collaborated and examined educational, social, and cultural factors as well as intervention programs that had been successful in closing retention, achievement, and/or choice gaps between boys and girls. They were able to firmly conclude that classroom environments could affect patterns of girls' interest, attitudes, and achievement in science.

As a NISE scholar, Jane worked with Norm Webb, who challenged her to develop a metric for measuring progress toward equity. Her research into the national databases indicated specific grade levels and/or ages when specific educational conditions or challenges had to be met in order to ensure that a student could continue on a science/math pathway. The metric enabled researchers to measure the efficacy of various reforms, including the systemic initiatives, in reaching equity. It also was useful for teachers, administrators, and others involved in studying classrooms and schools' movement toward equity (Kahle, 1998a). The metric, as well as important research by others who studied equity within the context of reform, was published in a special issue of the *Journal of Women and Minorities in Science and Engineering* (Kahle, 1998a).

During this time, Jane also connected two aspects of her career—the early research with African American students in Gary, Indiana, and Chicago, Illinois—with the later study of systemic reform with African American students in Cincinnati, Dayton, Toledo, Columbus, Cleveland, and Akron, Ohio. As gender research progressed, it became increasingly obvious to Jane that "one size did not fit all girls;" that is, the factors affecting Latinas varied from those affecting Asian American or African American girls. Jane, her colleague Judith Meece, and former student, Kate Scantlebury, analyzed achievement, attitudinal, and behavioral data from *Discovery*. They found that attitudes as well as peer participation and home support correlated with science achievement differently for African American girls and boys (Kahle et al., 2000). Achievement for African American girls, compared to boys, was more strongly correlated to higher ratings of home support and peer participation; while for boys, achievement was more strongly correlated with positive attitudes toward science than it was for girls.

In 2002 Kahle won the American Education Research Association's (AERA) Willystine Goodsell Award for research about and advocacy for women. Her acceptance address, presented in 2003, examined the impact of possible test bias on the science achievement scores of girls and boys; for example, how content emphasis in standardized tests, such as NAEP and TIMMS, affects the scores of boys and girls differently. Jane's work in systemic reform, as well as her prior experience on national committees that set and examined science education policy, led her to critique the *No Child Left Behind* policy, which required disaggregation of scores by many subgroups (ethnic/racial, geographical, historical, economical) in her acceptance speech. She examined the effect that changes in standardized test items had on girls and boys' scores in science from the 1969–1970 NAEP through 1999–2000. In addition, she examined how the format and type of items on the Programme for International Student Assessment (PISA) contributed to higher scores for girls, compared with boys—changes that essentially eliminated the gender gap favoring boys. Her analysis raised questions about whether the *No Child Left Behind* policy ignored gender issues by not requiring states to report students' achievement statistics by gender or by gender and race. Her address posed the question, "Will Girls Be Left Behind?" In a further effort to engage science educators with this question, Jane published her talk as a guest editorial in the *Journal of Research in Science Teaching* (Kahle, 2004).

CONCLUSION

One of Jane's greatest legacies is that a woman from a fiscally solvent, well-educated family can be invested in equitable education and making a difference in the lives that are so much less fortunate than her own. As the authors can attest, her accomplishments reflect more than a genuine interest, but decisive commitment and action.

The outcomes of Jane's work continue to inform reform efforts purposed with improving science and mathematics education. Contemporary reform efforts include accrediting bodies for teacher education requiring teacher education programs to gather and provide evidence of how their structures, activities, and candidate competencies are impacting student achievement. At the same time school districts and their teachers are faced with the challenges of providing evidence of their impact on student learning and achievement. In the face of these challenges that accompany large-scale and local reform efforts, Jane has prepared the groundwork that has yielded some insights into the manner in which the science education community can determine the impact that vision, relationships, and tenacity of scholarly work can have on individual student performances. Her scholarly pursuits and achievements provide important lessons and insights for researchers and educators who recognize the importance of addressing gender, diversity, and equity in such reform efforts.

More broadly, Jane's story and contributions provide important implications for those who dare to embrace life-changing transformations and etch out their own career curves. In these ways, Jane Butler Kahle certainly is a pioneer.

REFERENCES

Belland, J. C. (1991). Developing connoisseurship in educational technology. In D. Hlynka & J. C. Belland (Eds.), *Paradigms regained: The uses of illuminative, semiotic, and post-modern*. Englewood Cliffs, NJ: Educational Technology.

BSCS. (1963). *The biology teachers' handbook*. New York, NY: Wiley.

Carnes, G. N., & Damnjanovic, A. (1993). *A quantitative and qualitative study of inserving teachers to do inquiry-teaching*. A paper presented at the annual meeting of the National Association for Research in Science Teaching, Atlanta, GA.

Eccles, J. (1983). Expectancies, values, and academic behaviors. In J. T. Spence (Ed.), *Achievement and achievement motives* (pp. 75–146). San Francisco, CA: Freeman.

Eccles, J. (1989). Bringing young women to mathematics and science. In M. Crawford & M. Gentry (Eds.), *Gender and thought: Psychological perspectives* (pp. 36–58). New York, NY: Springer-Verlag.

Fennema, E. (1981). Women and mathematics: Does research matter? *Journal for Research in Mathematics Education, 12*, 380–385.

Fennema, E., & Sherman, J. A. (1978). Sex-related differences in mathematics achievement, spacial visualization, and related factors: A further study. *Journal for Research in Mathematics Education, 9*, 189–203.

Holliday, W. G. (2003). Influential research in science teaching: 1963–Present [Special Issue]. *Journal of Research in Science Teaching, 40*(Suppl.), v–x.

Horizon Research, Inc. (HRI). (1994). *Ohio's statewide systemic initiative Project Discovery: Classroom evaluation report –1994*. Chapel Hill, NC: Author.

Horizon Research, Inc. (HRI). (1995). *Ohio's statewide systemic initiative: Impact highlights: Evaluation summary*. Chapel Hill, NC: Author.

Horizon Research, Inc. (HRI). (1996). *The impact over time of Project Discovery on teachers' attitudes, preparation, and teaching practice*. Chapel Hill, NC: Author.

Kahle, J. B. (1982). *Double dilemma: Minorities and women in science education*. West Lafayette, IN: Purdue University Press.

Kahle, J. B. (Ed.). (1983, October). *Girls in school: Women in science*. Washington, DC: National Science Board, Commission on Pre-college Education in Mathematics, Science and Technology. (ERIC Document Reproduction Service No. ED258–812)

Kahle, J. B. (1998a). Equitable systemic reform in science and mathematics: Assessing progress. *Journal of Women and Minorities in Science and Engineering, 4*(2/3) 91–112.

Kahle, J. B. (1998b). *Reaching equity in systemic reform: How do we assess progress and problems?* (Research Monograph No. 9), Madison: University of Wisconsin, National Institute for Science Education.

Kahle, J. B. (2004). Will girls be left behind? Gender differences and accountability [Guest editorial]. *Journal of Research in Science Teaching, 41*, 961–969.

Kahle, J. B. (2007). The road to reform: A personal journey. In K. Tobin & W.-M. Roth (Eds.), *The culture of science education: Its history in person*. Rotterdam, The Netherlands: Sense.

Kahle, J. B., & Lakes, M. (1983). The myth of equality in science classrooms. *Journal of Research in Science Teaching, 20*, 131–140.

Kahle, J. B., & Meece, J. (1994). Research on gender issues in the classroom. In D. Gabel (Ed.), *Handbook of research in science teaching and learning* (pp. 542–557). Washington, DC: National Science Teachers Association.

Kahle, J. B., Meece, J., & Scantlebury, K. (2000). Urban African American middle school science students: Does standards-based teaching make a difference? *Journal of Research in Science Teaching, 37*(9), 1019–1041.

Kahle, J. B., & Nordland, F. H. with Charles, H., & Sims, M. (1974). An alternative education experience: A workable program for science in large-city schools for the disadvantaged. *The Science Teacher, 41*,(9), 44–49.

Kahle, J. B., Parker, L., Rennie, L. J., & Riley, D. (1993). Gender differences in science education: Building a model. *Educational Psychologist, 28*(4), 379–404.

Kahle, J. B., & Rogg, S. R. (1996). *A pocket panorama of the Landscape Study, 1995* [Booklet]. Oxford, OH: Miami University.

Kahle, J. B., & Rogg, S. R. (1997). Assessing systemic change: Ohio's Statewide Systemic Initiative, Discovery. *Systemic Initiatives, 1*(3), 10–11.

Kahle, J. B., & Woodruff, S. B. (2011). Science teacher education research and policy: Are they connected? In G. DeBoer (Ed.), *Research in science education: Vol. 5, The role of public policy in K–12 science education* (pp. 47–75). Greenwich, CT: Information Age.

McDermott, L.C., & Physics Education Group (1995). *Physics by inquiry: An introduction to physics and the physical sciences* (2 Volume Set). Hoboken, NJ: Wiley.

Phelps, M. E. (Ed.). (2004). PET: Molecular imaging and its biological applications. New York, NY: Springer.

Rowe, M. B. (1986). Wait time: Slowing down may be a way of speeding up! *Journal of Teacher Education, 37*(1), 43–50.

Tobin, K., Kahle, J. B., & Fraser, B. J. (Eds.). (1990). *Windows into science classrooms: Problems associated with higher-level cognitive learning.* New York, NY: Falmer.

CHAPTER 5

ROBERT E. YAGER

Science/Technology/Society (STS) Protagonist

Barbara S. Spector
University of South Florida

Megan Ehlers
Sarasota High School

Alfred Goytia
University of South Florida

Dr. Robert E. Yager is a legend in his own time. As a lifelong resident of small town Iowa, he has affected science education across the United States and worldwide. He is the most prolific author in science education (more than 500 books, monographs, and research papers), has the longest STS lineage with more than 100 doctoral degrees, and an STS advocate par excellence. Even though he was not the first science educator to use social issues to teach science, his impact on K–12 science teachers is unparalleled. His message has been, and still is, clear: Science teachers must change to

Going Back for Our Future II, pages 87–97

active constructivist instruction using social issues as a vehicle to make K–12 science learning attractive, relevant, meaningful, effective, and long-lasting for students.

As a leader in the science education enterprise, Dr. Yager is exceedingly approachable and welcoming. His humble stance may well contribute to his success. Even though credit is publicly given to him for the success of many STS initiatives, he is quick to highlight numerous colleagues and students who worked with him to forward STS and the opportunities to pioneer STS he has had by leading numerous professional organizations: As research director, then president of National Science Teachers Association (NSTA), and later chair of NSTA's Horizons committee he was able to guide the organization's resources (power) toward STS initiatives, such as the Chautauqua Program (1983) Search for Excellence (1983), Scope, Sequence and Coordination (SS&C) (1994), and the Exemplary Science Programs search (2005).

"It is all about what learners are able and willing to *do* with the science education available to them." Although I don't remember ever hearing Robert Yager say these exact words, that sentence sums up for me what gave him inspiration and direction (R. Yager, personal communication, 2012). A key feature of his work, distinguishing it from much of the earlier thinking by others addressing the need for improvement in science education, was his attention to students' learning, in contrast to curriculum used for teaching. Change initiatives for the science education enterprise prior to 1979 assumed changing the curriculum programs used by teachers would cure the ills of the enterprise. The alphabet curricula of the 1960s and early 1970s, sometimes described as intending to be "teacher proof," stand out as examples of large-scale funding curriculum initiatives that left the enterprise still in need of change.

Yager emphasized outcomes (goals) for students' learning and *how* students are empowered to learn, in order to function effectively in this scientifically and technologically driven democratic society. Yager's perspective was influenced by Dr. Paul DeHart Hurd, the "grandfather" of science education. Yager had additional perspectives that were uncommon when he began what might be described as his "STS crusade," including the following: He had the strong desire to do research that would be immediately applicable by teachers. He believed educating teachers over an extended period of time, as opposed to short workshops, was more critical than just creating new curriculum programs for teachers to use with their students. Honoring teachers' input through large-scale collaborative partnering with teachers and schools was essential. Input from parents, community members, and scientists was important. He was zealous about conducting and publishing research on every project initiated to expand understanding of teacher education.

During the 1970s, Yager was influenced by ongoing work in Europe in which science, technology, and society were being combined. That work stimulated reflection on his own experience in the University Laboratory School at the University of Iowa: He and a social studies colleague established a course labeled, *Science and Culture.* He recognized that society's growth and sustainability required applying science to social problems. This was long before today's sustainability focus became popular in the United States. Solving human problems is often labeled technology and/ or engineering. Developing both hard and soft technologies radically accelerates social change. The natural blend of science and technology with society became the ongoing theme of his work.

Establishing a Reform Agenda

The National Science Foundation awarded a grant to Dr. Norris Harms in 1977 titled, *Project Synthesis.* The goal was to ascertain the desired state for science education in the United States and to design a reform agenda to attain that state. The grant work began with reinterpreting the work of three prior NSF status studies: Stake and Easely (1978); Weiss (1978); Helgeson, Blosser, and Howe (1977) and the Third National Assessment of Educational Progress (U.S. Department of Education (1978). Those studies indicated extant goals for science were typically success in the next grade, getting into college, or becoming professional scientists. As one of the architects of Project Synthesis, Yager promoted goals that went beyond those existing and grouped them together under the following labels: meeting personal needs, societal issues, academic preparation, and science and technology career awareness.

Yager stated in one of the Project Synthesis committee reports what he referred to as the "unique" goals of science education: increasing intellectual development, teaching skills necessary for application and understanding, having students experience hands-on science, developing attitudes and values toward science and technology, and the consideration of science as part of an individual's basic skills (Yager, 1978). Thus, goals were revised from a focus on content to only those that would lead to developing a scientifically and technologically literate and capable citizenry. Strategies for teachers to use to attain the new goals and information enabling teachers to understand the status of science teaching in K–12 classrooms were included in the report.

Project Synthesis required several millions of dollars in funding, produced thousands of pages of information, and brought Yager to the realization that biology in school programs could be summarized in one word: textbooks. The major determinate for all content taught in schools was the standard science textbook (Harms & Yager, 1981). The studies also revealed that the methods used by classroom educators were exclusively classic didactic teaching, with the rare use of inquiry (Yager, 1983).

Yager concluded the reforms necessary to produce an exemplary classroom, illustrating that the desired state required an emphasis on changing science teachers' theoretical knowledge. In his view (contrary to views accepted by scientists and many policymakers), it was not sufficient for teachers to be fluent in the content; they needed to understand how to facilitate their students' mindset to promote scientific literacy. Yager emphasized the need for science educators to not only create curriculum, be effective teachers, and address academic preparation, but to also blend knowledge from other subject areas and apply it to science education. Project Synthesis illustrated this by using literature from the sciences, education, psychology, sociology, history of science, and the philosophy of science.

Next Steps

With Project Synthesis goals explicated, Yager continued further developing a variety of teacher education initiatives at the University of Iowa for preservice and in-service teachers to learn ways to attain the desired state. Teachers in other parts of the United States benefited from the Iowa experiences when the Chautauqua Program was funded in 1983. In the Chautauqua program, teachers were educated and observed for the period of a year. Yager recognized that professional development for educators had no staying power without follow-up. For successful implementation of a program, it was necessary for a researcher to follow-up with the teachers to support and verify progress, indicating positive results beyond the training period.

The desired outcomes emerging from Project Synthesis were then used, under Yager's leadership, as criteria in the NSTA *Search for Excellence* (1983) to find instances in the United States where these outcomes and facilitating processes actually functioned and demonstrated success. All the goal clusters were deemed necessary for a program to be exemplary. The search revealed a variety of factors that the excellent schools had in common in addition to meeting the criteria in the goal clusters. These factors follow: Their excellent "design" could be traced to one person. They developed local curricula and adherence to a textbook was limited. There was an increase in attention, time, funds, and dedicated staff for curriculum development, leadership, and in-service responsibilities. They had strong community involvement and ties to a variety of colleges and universities. Applying constructivist learning theory as a framework for analysis to the interplay of these factors can help to explain the role of many of these factors in the exemplary programs.

The science education enterprise began attending to the implications of constructivism for learning science in the 1960s, and it became common in the 1970s (Lockhead & Yager, 1996). Constructivist learning combined seamlessly with project-based learning and fit with STS. This combination allowed for an increase in the positive attitudes students experienced during science learning. Gain in positive attitude was a stark contrast to the standard in which a large percentage of students' interest in science learning declined, especially during the middle school years (Yager & Yager, 1985). The increase in personal relevance and project-based learning conveniently fit into recent science reform efforts (Seok & Yager, 2004). Several factors such as making science relevant and interesting to students, giving students hands-on experience via project-based lessons, involving the community in science learning, and developing teacher pedagogy to include constructivist methodology crescendoed in the Science, Technology and Society movement. The emergent characteristics from the Iowa group's ongoing research were infused into Yager's future teacher education initiatives.

The NSF funded an SS&C teacher education project in 1994, administered through the NSTA. Six states were involved, each using its own approach to attain the common goals of the grant. Yager led the Iowa initiative using STS as the driving force for the teacher training and curriculum materials developed. Research indicated the Iowa SS&C teachers improve their confidence, increased their understanding, engaged in action research, and implemented constructivist approaches when teaching. By not only handing teachers curriculum, but by educating them on its use, Yager laid the foundation to impact student-teacher interactions far into the future.

Outcomes for students of the Iowa SS&C teachers were examined in six domains: concept, process, creativity, attitude, application, and worldview. They outperformed the control group in all domains. Especially noteworthy were the outcomes for lower level and female learners regarding attitude. They traditionally lacked interest in science. This illustrates the way Yager worked at the front line to change science education with the teachers and their students.

As the National Research Council began development of the *National Science Education Standards* (NSES), published in 1996, Yager worked tirelessly to ensure the principles of STS became embedded in the standards. Findings from his work and that of his many doctoral students produced convincing research that led to success in getting key STS concepts embedded, but without the label STS. As a result, some other STS advocates thought it might be unnecessary to even continue explicit STS initiatives by name. As the new millennium progressed, Yager sought to find programs in schools actually implementing NSES. He led an NSTA initiative resulting in publication of a series of books ascertaining the nation's progress toward NSES titled, *Exemplary Science Programs.* Nine books in the series were published between 2005 and 2012.

Where It Began

Dr. Yager's interest in teaching was influenced by his mother having been a teacher. In middle school he decided teaching was to be his career. He excelled in many disciplines as a student throughout his academic career, but found science easier than the others. He did however graduate from college in three calendar years with multiple majors: science, English (and drama), and government. His first teaching position in 1950 was in both English and science in a small Iowa secondary school. During the Korean War, draftee Yager taught English and arithmetic in Germany to soldiers with 4th-grade competencies. What he learned about the difficulties these learners had reinforced his original career decision and triggered an interest in how it was that such difficulties might be overcome. Back in the United States, he went to the University of Iowa for doctoral studies. The intake advisor with whom he met was a plant physiologist who encouraged Yager to earn his PhD in the same area. To the good fortune of the science education enterprise, the narrowness of the life of a researcher was not as appealing as life as a teacher. The University of Iowa invited Yager to be Acting Head of Science Education and Head of the Science Department at University High School in 1956. He taught eighth graders in this laboratory school, which closed in 1972. A turning point in his life occurred when the school began doing research on learning and the environment. In addition, teaching the *Science and Culture* course accepted for both science and social studies credit enabled him to obtain his first federal grant. So began his path as a grants man, receiving and directing more than 100 NSF-funded projects plus funds from other sources. It also heralded the integration of different disciplines into the science/technology/society (STS) reform movement.

Looking Back at Societal Context

Science education research was in its infancy in the late 1950s when the field of science itself became a focal point in the United States due to "Sputnik Shock" in 1957. Later this phenomenon resurfaced in the 1980s as "Japan Shock" (Yager, 2000a). As the United States entered the global economy, scientists, government, and politicians realized the need to prepare students and teachers to better compete with other countries. This necessitated an introspective examination of science education in the United States. There were very few major efforts to educate science teachers at this time, a weakness that became a dominant focus of Yager's research (R. Yager, personal communication, 2012).

The relatively small amount of research in science education that had been published before the 1960s was attempting to mimic the "hard sciences." Science education was thoroughly entrenched in this paradigm, which restricted measurement to purely *a priori* quantitative methods. The science education researchers mirrored the situation of scientists who worked isolated from scholars in other fields and from the general public. Science was taught using exclusively didactic techniques. Yager identified these conditions as limitations to the well-being of the science education enterprise and worked tirelessly throughout his career to identify and offer viable and often superior alternatives.

Research Methods and Results

Early in his career, Yager's research methods were solely quantitative, consistent with accepted procedure in the sciences and in science education at that time. Given his scientific background, a PhD in plant physiology, this was predictable. In his initial publications, his research focused on the science laboratory and its influence on students' ability to learn scientific concepts. During his initial investigations, he realized there were interesting discrepancies between what he observed happening in the laboratory and the results of his quantitative measurements. Yager's position in the Laboratory School gave him easy access to students and teachers, thus facilitating his research. The opportunity to teach students and observe and interact with other teachers in the classroom helped him foster an uncommon perspective for the late 1950s and 1960s, seeking education research findings that could be put into practice almost immediately.

Yager's first publications were straightforward in nature. This was evident in an early publication involving research into whether or not 9th- and 10th-grade students were able to handle the biology classroom course material (Yager, 1962). The article is short and quantitative. His mindset parallels the manner in which first year teachers face challenges by focusing on details that can be easily rectified.

Yager's research on teacher education evolved in the next two decades aided by the large amount of financial support which, in the 1960s, the NSF dedicated solely to the study of science and mathematics education across the United States. The grants contributed to the University of Iowa being responsible for 90% of the research in science education in the country at the time and provided Yager the opportunity to branch out and study science education in a broader sense, not solely in the classroom. Survey data collected from 1960–1980 were used to create a report, *Status Study of Graduate Science Education in the United States, 1960–1980* (1980), analyzing the state of science education in graduate programs across the country.

Yager found that within the period of 1960–1970, graduate institutions in science education nearly tripled. The number of grants designated for science education increased in the middle of the 1970s until suddenly the trends stopped and regressed. Money in the form of grants for teacher inservice, research, and curriculum development dropped sharply by 1980 (Yager, 1980).

Dr. Yager's future research methods were influenced by his experience with the Project Synthesis team using a discrepancy model, wherein the description of the desired state was followed by descriptions of the actual state of affairs. The information analyzed during Project Synthesis consisted of a wide variety of writings and reports concerned with current projects, viewpoints, and research. Project Synthesis, which was viewed as a qualitative and normative research procedure, was in itself just as valid as traditional quantitative analysis (Yager 1980).

Years earlier, when Yager had employed quantitative analysis in his biology laboratory studies, he had reported that "confidence in the statistical treatment provided considerable interest in terms of the implications of his results" (Yager, Engen, & Snider, 1969). In the paper, "Effects of the Laboratory and Demonstration Methods Upon the Outcomes of Instruction in Secondary Biology," Yager seems to be in the initial stages of becoming aware of the discrepancies between what he was observing in the classroom and his quantitative measurement. Yager then proceeds to recognize the issue of validity and reliability in currently used measurement instruments in the social sciences, and proposed qualitative research as an alternative to the current norm, which consisted of quantitative analysis. The needs described for research in science education in the article, despite being written over 45 years ago, still mirror current issues in educational research.

In the 1980s, Yager characterized qualitative synthesis as consisting of diverse research teams, providing multiple sources of data (mostly qualitative) and including one or more conceptual schemes. In "Factors Involved with Qualitative Synthesis: A New Focus for Research in Science Education" (1982), Yager referred to Project Synthesis as an example in which qualitative synthesis provided a superior analysis of the state of science education to that of the more traditional quantitative analysis. He noted that qualitative synthesis results in a new model to be used for further research and action recommendations (Yager 1982). By emphasizing qualitative data from multiple sources, as well as approaches, the result was a comprehensive plan for the science educator. This article and Project Synthesis are an example of how Yager wished to directly impact science education, not just report on its current state (Yager, 1982).

There have been science educators during Yager's career who have disputed the concept of STS for school science. They believe the silo of science is sacred and should not be joined with technology or social science, despite Yager's sagacity, innovativeness, perseverance, and database of evidence. He has been able to refute their attacks, occasionally using their own evidence to support his position (Spector & Yager, 2011).

"Scope, Sequence and Coordination" enacted from 1990 through 1997 (Yager, 1999) was a case in point, where the funding for his work came from the NSF through the NSTA whose executive director and principal investigator for the grant did not accept STS. The executive director went so far as to leave Yager's Iowa SS&C out of the second round of funding from NSF. This turned out to be fortuitous, because it provided a dataset to contrast accomplishments with Iowa's SS&C.

In 2008, he once again argued for STS in science classrooms. Two middle school science classrooms were taught using the Science Technology and Society approach, and the other using the textbook as the driving force (Yager & Akcay, 2008). Middle school was described as a key grade level for this. The two teachers chosen for this research project were experienced, and both were concerned about the waning interest of the students they were teaching. Instead of just writing about the philosophy behind the STS approach in science education in this article, it was put to practice.

The results of this mixed-methods (quantitative and qualitative methods) study indicated there was no basis for STS adversaries' claims that students would not learn as much basic science as in a traditional classroom. STS students were as successful as the textbook-driven students at learning basic science concepts. STS students, however, outperformed the textbook-oriented students in developing positive attitudes toward science and being more actively engaged in discussion and overall other characteristics desired in an STS classroom.

More than 40 years after the push toward improvement in science education due to Sputnik and later Japan, once again global competition with other industrialized and emerging nations came to the fore on the political stage. The *National Science Education Standards* emerged, which recognized the need for a more scientifically literate society, emphasizing real-world experiences and integration of technology. When discussing the future of reform in science education, Yager stated, "These are interesting times. They will be even more interesting if we are able to maintain public support and interest in the current reform process (Yager, 2000a)." More than 12 years later, policymakers and implementers have become more interested in improvement in science education. It is however apparent their interpretation of reform centers on the use of standardized tests, a method in sharp contrast to Yager's proposals for reform (R. Yager, personal communication, 2012). Standardized test scores are labeled as inaccurate indicators of student performance, only measuring information in textbooks, which is not consistent with the vision put forth by the NSES. Yager noted current research needs to be put to use by science educators and administrators.

By citing results from his research, which demonstrates that using STS in the classroom produced student success with content equivalent to more traditional classes, however, provided value added by students' improved attitudes and creativity, Yager provides a pathway to reach the NSES vision. Altering the nature and use of high-stakes testing to address the six domains important to learning science offers possibilities to meet the political need for accountability (Yager, 2000b).

REFERENCES

Harms, N., & Yager, R. E. (Eds.). (1981). *What research says to the science teacher* (Vol. 3). Washington, DC: National Science Teachers Association.

Helgeson, S. L., Blosser, P. E., & Howe, R. W. (1977a). *What research says to the science teacher* (Vol. 3). Washington, DC: National Science Teachers Association.

Lockhead, J., & Yager, R. E. (1996). Is science sinking in a sea of knowledge? A theory of conceptual drift, In Yager R.E. (Ed.) *Science/technology/society as reform in science education.* Albany: State University of New York.

Seok, P. O., & Yager, R. F. (2004). Development of constructivist science classrooms and changes in student attitudes toward science learning. *Science Education International, 15*(2), 105–113.

Spector, B., & Yager, R. E. (2011). The many faces of STS social issues in education. In S. Totten & J. E. Pederson (Eds.), *Teaching and studying social issues: Major programs and approaches* (pp. 277–311). Charlotte, NC: Information Age.

Stake, R. E., & Easley, J. (1978). *Case studies in education, Vols. I & II.* Washington, DC: U.S. Government Printing Office, Center for Instructional Research and Curriculum Evaluation, University of Illinois at Urbana-Champaign.

U.S. Department of Education (1978). Institute of Education Sciences, National Center for Education Statistics, Third National Assessment of Educational Progress (NAEP), Science Assessments. Washington, DC: Author.

Weiss, I. R. (1978). *Report of the 1977 National Survey of Science, Mathematics, and Social Studies Education: Center for Educational Research and Evaluation.* Washington, DC: U.S. Government Printing Office.

Yager, R. E. (1962). Variation in the success of ninth and tenth year students in general biology. *The American Biology Teacher, 24*(8), 569–572

Yager, R. E. (1978). Priorities for research in science education: A study committee report. *Journal of Research in Science Teaching, 15*(2), 99–107

Yager, R. E. (1980). *Status study of graduate science education in the United States, 1960–1980.* Final report for NSF Contract #79-SP-0698.

Yager, R. E. (1982). Factors involved with qualitative synthesis: A new focus for research in science education. *Journal of Research in Science Teaching, 19*(5), 337–350.

Yager, R. E. (1983). The crisis in biology education. *Elements, 1,* 14–27.

Yager, R. E. (1999). Scope, sequence and coordination: The Iowa project, a national reform effort in the USA. *International Journal of Science Education, 21*(2), 169–194

Yager, R. E. (2000a). The history and future of science education reform. *Clearing House, 74*(1), 51–54.

Yager, R. E. (2000b). A vision for what science education should be like for the first 25 years of a new millennium. *School Science and Mathematics, 100*(6), 327–341.

Yager, R. E., & Akcay, H. (2008). Comparison of student learning outcomes in middle school science classes with an STS approach and a typical textbook dominated approach. *Research in Middle Level Education Online, 31*(7), 1–16.

Yager, R. E., Engen, H., & Snider, C. F. (1969). Effects of the laboratory and demonstration methods upon the outcomes of instruction in secondary biology. *Journal of Research in Science Teaching, 6,* 76–86.

Yager, R. E., & Yager, S. O. (1985). The effect of schooling upon understanding of selected science terms. *Journal of Research in Science Teaching, 22*(4), 359–364.

CHAPTER 6

GEORGE WILLIAM JEFFERS

Dedicated Teacher, Scholar, and Change Agent

Beth Shiner Klein
SUNY Cortland

Starlin D. Weaver
Salisbury University

Dr. George William Jeffers (1897–1995) was a science teacher educator dedicated to teachers colleges and proud of his role in preparing primarily female students at one of the oldest women's colleges in the nation. A founder of the National Association of Biology Teachers and charter member of the National Science Teachers Association, he was so respected in his field that he was one of only four people invited to advise the U. S. Commissioner of Education on Science Education in 1944. He was so loved by his students that they wrote poems about him and continued to write to him to ask for professional advice 35 years after he left his classroom. A look at his writings, letters from colleagues, and students reveals a complex thinker, an inspirational and giving teacher, a gentle soul, a change agent, and a highly respected and humble scholar.

Going Back for Our Future II, pages 99–107

George William Jeffers was born in 1897 in Freshwater, Caronear, Newfoundland, Canada. Upon graduating from a teacher preparation program at Methodist College, St. John's, Newfoundland, Dr. Jeffers spent 4 years teaching in the Newfoundland schools. He later attended Boston University, graduating in 1924 with a Bachelor of Science degree in biology and in 1925 with a master's degree in biology. Following the completion of his master's, Dr. Jeffers went to the University of Toronto and worked as a biology research assistant and began to work toward his doctoral degree. In 1926, he joined the staff of the State Teachers College (later known as Longwood College and now Longwood University) in Farmville, Virginia. He continued to work on his PhD in biology, and while on sabbatical in 1931 he completed his degree from the University of Toronto. His dissertation focused on the life history of the capelin, a small fish in the smelt family (Archives Canada, n.d.).

During Dr. Jeffers' years at Longwood College, he served as Chair of the Natural Sciences Department. Upon his retirement, Longwood named a new auditorium after him. In addition, in 1974 he was awarded an honorary doctorate of science from the Memorial University of Newfoundland.

A scholar in his own right, Dr. Jeffers was a Fellow of the American Association for the Advancement of Science (AAAS) and well known in his field of biology. He directed the Chesapeake Bay Fisheries Commission from 1943 to 1945. He was also a founding member of National Association of Biology Teachers (NABT), a charter member of National Science Teachers Association (NSTA), and started the Virginia Junior Academy of Science (VJAS).

FOUNDING OF THE NATIONAL ASSOCIATION OF BIOLOGY TEACHERS

Following a call to action on the need for improved biology teaching, a 1936 address by outgoing AAAS president Dr. Oscar Riddall announced the establishment of an AAAS committee on biology teaching. This committee led to a July 1, 1938, meeting in New York City of "eighteen enthusiastic biologists" of which Dr. Jeffers was one. They had the "avowed purpose of setting up a national association" (Jeffers & Lightner, 2001, p. 6). Jeffers described the scene:

> Rarely had the sun come up over that city's skyline with greater brilliance and with more determination to establish a record high temperature. Since such creature comforts as air conditioning were still in the future, the atmosphere within the conference room of the Hotel New Yorker matched the blistering heat outdoors. Only a transitory respite was possible with shedding of coats and ties, loosening of collars, and ingestion of ice water. However, matching

> the heat of the day was the enthusiasm with which the delegates welcomed the many-sided problem that had confronted them. (Jeffers & Lightner, 2001, p. 6)

Jeffers went on to express how he and the rest of the delegates were charged with establishing a national organization devoted to the teaching of biology and that the National Association of Biology Teachers (NABT) would be "under the control of high school teachers of biology." In that meeting, they established the NABT, drafted a constitution and bylaws, and established guidelines for a brand new journal, which first appeared in print only 90 days later (Jeffers & Lightner, 2001).

At that initial meeting, Jeffers was elected first vice president. One year later in June of 1939, the fledgling NABT had 1,600 members. Dr. Jeffers was program chair of the 1939 NABT convention in which, in addition to a full program, he arranged a "rather unique banquet featuring major speakers, namely Dr. William Weston of Harvard and Dr. E.G. Conklin of Princeton" (Jeffers & Lightner, 2001, p. 12). Also in 1939, with a national election, Dr. Jeffers was chosen as president-elect of the organization. Jeffers took his term as NABT president 1940–1941, serving as the third president of the organization and the first to represent college-level biology teaching. When commenting on the turnover of officers, Jeffers said, "No better evidence could be advanced in support of the contention that the National Association of Biology Teachers belonged to the rank and file biology teacher, and was not in the control of any clique, however wise and benevolent" (Jeffers and Lightner, 2001, p. 12).

During Jeffers' tenure as president, the NABT added 294 new members, its "financial picture improved," and they added six more affiliate organizations, bringing the total to 19 (Jeffers & Lightner, 2001, p. 12). This was all taking place as the country watched what would become World War II escalate in Europe and in the Pacific. Like many associations, due to the onset of World War II, NABT lost revenue and membership.

In 1944, the National Science Teachers Association (NSTA) was created as a merger of the American Science Teachers Association and the American Council on Science Teaching (a National Education Association component). Dr. Jeffers was also involved in this beginning and is listed as a charter member of the NSTA. Interestingly, in the late 1940s, Dr. Jeffers found himself in bit of a controversy between NABT and NSTA. His friend and colleague Dr. Ephraim Laurence Palmer of Cornell University was president of NABT in 1947 at the time the NABT was receiving pressure to become a subgroup of the NSTA.

Although it would make sense that the NSTA would speak for several science teaching organizations, the education section of the American Chemical Society and the American Association of Physics Teachers chose not affiliate with the NSTA at that time. According to Jeffers in a personal communication from Palmer, Palmer "kept NABT from being steamrolled into NSTA. It was nip and tuck for a while as to whether we would just be the biology department of NSTA" (Jeffers & Lightner, 2001, p. 17).

In a letter from Palmer to Jeffers dated December 1, 1946, Palmer indicated that Jeffers did not feel NABT should be a part of NSTA, "There has been a long story of high pressure influence on the part of the N.S.T.A. to dominate the science teaching field." To this day, the NABT is still a separate organization from the NSTA.

ROLE IN THE VIRGINIA ACADEMY OF SCIENCE

Dr. Jeffers went directly from serving as NABT president to serving as president of the Virginia Academy of Science (VAS) from 1941 to 1942. During that time period the academy started the *Virginia Journal of Science*, an outgrowth of an earlier journal, *Clatonia*. Dr. Jeffers and other active members of the VAS also focused their attention on the science opportunities for youth of Virginia.

In 1940, Jeffers proposed, along with R. C. Berry, the Virginia State Chemist, a VAS resolution for the "formation of a full-fledged junior academy." This resolution was adopted and it became the incoming president's responsibility to form a committee to establish the Virginia Junior Academy of Science. Dr. Jeffers was that incoming president.

Jeffers' presidency came at an interesting time in history. As shared by Webb (2001),

> December 7, 1941—six months into George Jeffers' term as President of the Virginia Academy of Science—the Japanese bombed Pearl Harbor. One day later the United States Congress formally declared a state of emergency, dramatically shifting the careers of many Virginia scientists. Unlike the First World War, when scientists left their academic or (in a few cases) industrial posts to serve their country overseas, World War II was fought inside the walls of state and federal laboratories as well. (p. 63)

As such, VAS members were encouraged to register in the National Roster of Scientific and Specialized Personnel to assist the war effort.

In addition to focusing on the development of the Virginia Junior Academy of Science, Jeffers was instrumental in encouraging science clubs in schools throughout Virginia. Dr. Jeffers, along with other VAS members, gave several presentations and radio addresses to promote science clubs in Virginia schools. In addition, in November 1939, Jeffers published an article in the *Virginia Journal of Education*, titled "Science Clubs in the Schools" (Heatwole, n.d.). Continuing the campaign, Jeffers, along with Dr. Sidney Negus, visited numerous school districts speaking on science clubs. Some

26 of these clubs assisted in helping to form the VJAS (Heatwole, n.d.). Although VAS finances were tight, as Jeffers left office, he led the Academy in developing "a substantial budget for the support of the work of the Junior Academy of Science for another year, and that the Senior Academy thought it wise to have the next meeting of the Junior Academy in the same building with the Senior Academy to create more opportunity for discussions between the two groups" (Heatwole, n.d.).

According to Flory (1968), when discussing the history of the Virginia Academy of Science, "It is evident that Dr. George Jeffers has played a large and significant role in making the present history possible" (pp. 65–66). Flory also noted when referring to Jeffers' unpublished manuscript on VAS history "Dr. Jeffers manuscript carries a feeling of the 'life' of the Academy, and also contains much detail concerning the early meetings and actions of the academy"(p. 65).

The VAS membership noted Jeffers' contributions, awarding him an honorary membership in the association and their Ivey F. Lewis Distinguished Service Award. Dr. Jeffers is also memorialized by the VAS with the Dr. George William Jeffers Memorial Lecture given each year at the association's annual conference.

ADVISING THE U.S. COMMISSIONER OF EDUCATION ON SCIENCE EDUCATION

In 1944, Dr. Jeffers, along with Gerald Craig, Emil Massey, and Philip Johnson, were invited to advise the U.S. Commissioner of Education and his staff as to what services his office should provide science education. A portion of the introduction of their report was telling of the context of their work:

> Tragic as the present world conflict is, it has forced us as a nation to evaluate ourselves—our deficiencies and our capabilities. For example, woefully unprepared as we were at the time of Pearl Harbor, we have in less than three years amazed not only ourselves, but the world as to our productive genius. This ability to change rapidly from peace to a war-time footing has been in large part due to the mobilization of American science and to our flexible system of democratic education. (Craig, Jeffers, Massey, & Johnson, 1944, p. 1)

The report made several recommendations for the U.S. Office of Education to improve science education. Recommendations included

- request for studies on health education (as it connects to science education) and physical science (as it connects to "machines of industry and war")
- stronger liaison between college and high school science

- providing teacher institutes and workshops that focus on studying local, state, national and international problems with the emphasis on the role of science teachers in helping to solve these problems to "render a real service to the youth our nation through the development of closer cooperation between the science teachers at all grade levels and also between the science teacher and various other educational and community groups."
- better connection with science and industry facilitating closer correlation between science and vocational departments. (Craig, Jeffers, Massey, & Johnson 1944 pp. 1–3)

The report also recommended the U.S. Office of Education study the following: ways to strengthen (a) in-service and preservice science teacher education, (b) science curriculum, (c) student teaching practices, (d) mechanics of instruction, and (e) integration of science with different subject areas "with the purpose that science may be better coordinated in a program of education that looks to the well-rounded development of children and youth" (Craig et al., 1944, p. 7).

Additional recommendations included expanding science museums for the use of schools, particularly in rural areas; educating through radio; developing stronger visual aids; and development of a comprehensive clearinghouse in the United States for research studies in science education. Finally they recommended a continuous science program from elementary school through college "so that the science instruction in public education may be conceived as a continuous program" (Craig et al., 1944, p. 10).

ADVOCATE FOR TEACHERS COLLEGES

A copy of a presentation Dr. Jeffers gave at the April 1948 Conference on the Education of Teachers in Science (now called the Association for Science Teacher Education) revealed his view on the regional responsibility of the teachers college in the field of science. He was disappointed that teachers colleges, also known as normal schools, were moving away from their focus on teacher education to emulate broader, large universities. "Therefore, the name 'Normal School' was sloughed off, and most of them would today drop 'Teachers' from their name if they could. They would attain respectability even though it means becoming inept and isolationist" (Jeffers, 1948). He went on to explain how the state teachers colleges should look at state agricultural colleges that have "not forsaken its raison d'être, nor has it scorned 'the base degrees by which it did attain.' Teachers colleges could well emulate them" (Jeffers, 1948). He felt strongly that a teachers college should "steadfastly adhere to its original objectives and in so doing is in a fair way to attaining educational respectability, and in some instances at least, is rivaling the state university" (Jeffers, 1948). As predicted, today there are very few state institutions with "teacher" in still in their names, but there are still several with the "agriculture" in their title.

DR. JEFFERS, THE INSPIRATIONAL, GIVING TEACHER

Despite his active and time-consuming participation in science education in the national and statewide arenas as well as his scientific research, a review of multiple letters from students and dedications in Longwood College yearbooks indicate how much Dr. Jeffers gave of his time and talents to his students. As one biography noted about Dr. Jeffers,

> As with most great teachers, his greatest contributions and one in which he could take great pride, was his influence on his students. A man of sincerity and vivacity, one known for his sympathetic understanding, his infectious enthusiasm for biology was conveyed to student and colleague alike. Many of these students received advanced degrees and many became outstanding teachers of biology. Many of those students were impressed by his deep appreciation of literature, poems used to point up biological facts, and his attitude toward religion, demonstrated by a conviction that science and religion could be reconciled. (Virginia Junior Academy of Science, 1968)

For example, in a recollection, a 1957 graduate of Longwood wrote,

> I remember Dr. Jeff as a caring and dedicated professor who held a special place in the hearts of scores of his students. When I entered Longwood in the fall of 1953, I wanted to try and combine my elementary education major with biology. Those I talked with initially discouraged me. Eventually I was told I needed to talk with Dr. Jeffers, thinking, I'm sure, he would set me straight as to the foolishness of following through with my plan. When I explained my wish to Dr. Jeff, he said he had been looking for someone to do that for many years. He felt as I did, that young children needed to be involved in biology and outdoor education in a very early age and that they should be combined with many other subjects. He proceeded to discuss with me the path I needed to follow to accomplish my goal. I graduated with the equivalent of a biology major in addition to elementary education. I remained in touch with Dr. Jeff, as did many of his prior students, until his death. He was a very special person in the lives of many of Longwood's alumna and his positive influence has been felt within the educational system of Virginia and across the nation. It is not only his students that benefitted from their association with Dr. Jeff, but also the unnumbered elementary and secondary students who were taught by his graduates over the years. (G. P. Shiner, personal communication, August 2, 2013)

This former student's perspective on Dr. Jeffers' impact is supported by Dr. Sidney Negus of the Medical College of Virginia when, in 1961, he wrote about Dr. Jeffers' influence on science teaching.

> There hasn't been anyone in Virginia over more than a quarter of a century who has done more in encouraging inspirational teaching of biology, especially on the secondary level, than Doctor Jeffers. Few of us are privileged to multiply our teaching personalities to the exceptional degree that he had done over the years. Whenever one meets a former biology major from Longwood College, it's not long before Jeff's name is part of the conversation, and always most favorably so. (Negus, 1961)

And to highlight Dr. Jeffers' long-term impact, a letter dated March 2013 between two Longwood College alumni discussing their memories and how "Dr. Jeff" had inspired them, "Although I was an elementary major, I loved science and still take minerals to school when I sub at age 78. Can you believe that!"

CONCLUSION

Dr. Jeffers' contributions to the field of science education continue to live on. His work in establishing the National Association of Biology Teachers, National Science Teachers Association, Virginia Academy of Science, and Virginia Junior Academy of Science continue to influence science education to this day. Many of the recommendations made to the U.S. Office of Education are still timely today and can be found in most recent versions of the standards for teaching science, the Next Generation Science Standards (National Research Council, 2013).

Dr. Jeffers was proud that he taught at a state teachers college (primarily a women's college during his tenure) and was very proud of teaching that population of students. His connections with thousands of mothers and teachers have influenced generations of individuals to achieve scientific literacy. Upon his retirement in 1968, he wrote in a letter of appreciation for a recent accolade:

> I am all the more grateful to you, therefore, for having the ability to recognize that every once in a while at least one who would be recognized for minding his own business—and that the business of turning out, not scientists, but those who are to be so placed in society—as mothers or/and teachers that they can touch the lives of those more likely to become candidates for careers in science.

REFERENCES

Archives Canada. (n.d.). *Histoire administrative/notice biographique: George William Jeffers.* (Archive No. RCIA 206505) Retrieved from http://www.archives canada.ca/

Craig, G., Jeffers, G. W., Massey, E., & Johnson, P. (1944). *How the U.S. office of education can aid science instruction.* A Report to the U.S. Office of Education.

Flory, W. S. (1968). Credits, a history of the history. *The Virginia Journal of Science, 19*(1). Retrieved from http://www.vacadsci.org/history/credits.PDF

Heatwole, T. C. (n.d.). *Early history of the Virginia Academy of Sciences.* Special Collection, Virginia Tech.

Jeffers, G. W. (1948). *Regional responsibility of teachers colleges in the field of science.* Paper presented at the 1948 Conference on the Education of Teachers in Science. Washington, DC. George William Jeffers Special Collection, Greenwood Library, Longwood University.

Jeffers, G. W., & Lightner, J. P. (2001). *As we remember it: A personal recollection of the first four decades of the National Association of Biology Teachers.* Round Hill, VA: JLX.

National Research Council. (2013). *Next generation science standards: For states, by states.* Washington, DC: National Academies Press.

Negus, S. S. (September 7, 1961). Letter to the Virginia Academy of Science in support of Dr. Jeffers' Honor. Special Collections, Longwood University.

Virginia Junior Academy of Science. (1968). Biography of George William Jeffers recipient of the Ivey Lewis Distinguished Service Award. Virginia Junior Academy of Science Records, 1931–1999. Special Collections, Virginia Tech.

Webb, C. (2001). Linking allies and resources: Maintaining cohesion during World War II, 1940–1952. In *A history of the Virginia Academy of Science: 1923–2001.* Retrieved from http://www.vacadsci.org/history/Web/CWebb.html

CHAPTER 7

DR. NORMAN D. ANDERSON

Mentor, Science Teacher, and Author

J. Steve Oliver
University of Georgia

From the date of his birth until the early years of adulthood, Dr. Norman D. Anderson (Norm) identified rural northwest Iowa, a place commonly associated with pioneers, as home. Those pioneers were settlers with ambitions to put down roots and build a familial life based in rural agriculture. He heard stories of a migration by an earlier generation from Sweden to the Great Plains of Iowa. The stories told here are also of pioneering work but characterized in a different way. This chapter will focus on Norm Anderson's role as a pioneer who came to the profession of science education in the frenetic post-WWII years and was among those science educators who built the discipline that we now know. Like the other pioneers of this volume, Norm came to science education with the goal of improving the teaching and learning of science in our nation's schools.

On his family's farm in Freeman Township, Clay County, Iowa, Norm Anderson began his life in January of 1928. And his life always included education and science. If those labels were too formal, one might have said

Going Back for Our Future II, pages 109–127

that it included a great deal of study of natural phenomena with large doses of informal inquiry and problem solving thrown in. After a stint in the U.S. Army immediately after WWII, he graduated from the University of Iowa with a BA in general science. Soon afterwards his science teaching career was launched in Burlington, Iowa. While teaching, he completed his MA degree, also from the University of Iowa. Norm's early science teaching career would continue for a total of 7 years, although he had begun following the path of new opportunity while still teaching. In the late 1950s, Norm attended an NSF-funded academic year institute and began a doctoral program in science education at The Ohio State University under the guidance of Dr. John S. Richardson.

Norm finished his PhD in 1965 but had already begun his career as a faculty member in science education at North Carolina State University (NCSU) where he spent the remainder of his professional career. After promotion to the rank of Professor in 1971, he was named Alumni Distinguished Professor in 1973 and held this title until his retirement in 1994. After his retirement, he remained active in science education by serving as a developer of supplemental instructional materials for the SCI-LINK and GLOBE NET projects at NCSU.

Norm's career in science education was highlighted by the publication of 32 books, which range from middle school science textbooks to trade books (with titles like *Historical Catastrophes: Snowstorms and Avalanches*) to science instructional guides like *Investigating Science Using Your Whole Body.* Fully complementing his skills as an author, he published in 1981 a secondary science instructional methods book (co-authored with Dr. Ronald D. Simpson) titled, *Science, Students, and Schools: A Guide for the Middle and Secondary Teacher.* At the conclusion of his career, Norm published a series of sourcebooks (with a variety of co-authors, most commonly including H. S. Stubbs) through the SCI-LINK and GLOBE NET projects, including *Algae: A Sourcebook for Teaching About Harmful Algal Blooms*; *Acid Rain, Grades 4–8: A Sourcebook for Teaching About A Major Environmental Problem*; and *Lyme Disease: A Sourcebook for Teaching About a Major Environmental Problem.*

Among the many honors received during his distinguished career, Norm was twice named an Outstanding Teacher Award Winner by NCSU, was elected to the Academy of Outstanding Teachers at NCSU, was a visiting master teacher at Harvard University in 1964, elected a Fellow of the American Association for the Advancement of Science in 1969, elected to Phi Kappa Phi honorary society, received the Herman Gatling Award from the North Carolina Science Leadership Association in 2000, and was elected to the College of Education Hall of Fame at The Ohio State University in 2004. But having outlined the facts, let's go back to the story of science, science education, life on an Iowa farm, and the early life of Norman D. Anderson.

In the winter 1986–1987 issue of *The Rural Educator,* Norm wrote an article about his years of grade school education in a one-room schoolhouse in northwestern Iowa. Though increasingly rare today, one-room schools were, prior to WWII, frequently the site of formal education for the rural children of America. Norm's school was called Springdale School and was located in the center point of where four, 1-mile square (640 acres) sections of land met. In many rural areas of the United States, property boundaries were established by ridges, streams, big rocks, or other such geographic features. However, in the states of the Great Plains, the boundaries were laid out with a transit by a surveyor such that almost all boundaries were straight and also aligned to the cardinal points of a compass. His school, like so many others, served the grade school educative needs of those children from the farms on those four sections of land. A typical farm would occupy one "quarter-section" or 160 acres. The quarter-section of the Anderson family's farm was on the outer corner of the area serviced by the school, providing him a 2-mile walk one way between home and school. In the article from *The Rural Educator,* he described how on snowy days his father would take him, his siblings, and children from neighboring farms to school in a horse-drawn sleigh. For most of his time at Springdale School, he was the only child in his grade. The entire school population usually numbered a total of six or seven children in grades 1 through 8.

From Springdale School, he initially (in the fall of 1942) went to a small high school in Dickens, Iowa, which had about 25 students and more importantly used public school buses to transport children to and from the school site. However, after 2 years, the decision was made to close this school. In his second year, enrollment had dropped to about 18 students, with only two teachers. During this time, all of the resources of the country were directed to the war effort, and little was left for education of children. However, this seemingly unfortunate circumstance was actually a blessing, as he was able to attend a larger high school in Spencer, Iowa, for his last 2 years. He recalled this as "one of the best things that ever happened to me."

Spencer, Iowa, was an important town in northwest Iowa in the 1930s and 1940s. It had been drastically disfigured by a huge fire in June of 1931, but was reborn from the ashes to be a model downtown for small cities of the Midwest. The fire was the result of fireworks ignited in the basement of Bjornstad's Drugstore, where they were stored; but even this event was important to the professional life of Norm Anderson. In 1983, writing with his frequent co-author Walt Brown, Norm published a book titled *Fireworks! Pyrotechnics on Display,* which among other topics described the Spencer fire and its source in those stored fireworks.

High school in Spencer, Iowa, was a great experience for the young farmboy. Norm entered his senior year of high school just as WWII ended in the Pacific theater in the summer of 1945. The influence of the war was evident in the curriculum, which allowed him opportunities to take courses in aeronautics, radio, and chemistry. One teacher, Mr. Myhr, taught all of these, and Norm cites him as a powerful influence on his understanding of the subject matter as well as instilling in him an early understanding of how to teach those subjects. For instance, Norm always looked forward to Mondays as it was on this day that Mr. Myhr put a famous quotation on the top of the chalkboard. Norm would adopt the use of famous quotes in his teaching and pass this pedagogy along to prospective science teachers in later years as well. Seemingly mundane uses of basic instructional technologies, like the chalkboard, were put to innovative uses in these postwar school days. Mr. Myhr would have students go to the chalkboard to work problems, attempting to dispel some challenging aspect of a chemistry quandary. From this inspiration, the budding science teacher Norm reported he later realized that

> I could watch several students working on a problem and often spot where they were having difficulties. For example, students unable to write the correct formulas for sodium hydroxide and hydrochloric acid will not be able to write the correct equation showing how they react. (N. D. Anderson, personal communication, September 12, 2014)

Another teacher, who provided an important role model of the mentor who keeps up with his students subsequent to their school years, was Mr. Podol, whose subject was American government. Norm told the story of how he encountered Mr. Podol several years after graduating from high school. Norm was then a student at the University of Iowa, 250 miles to the southeast of Spencer. Mr. Podol had also gone to the University of Iowa for summer coursework and upon seeing Norm back in town told him of what he had learned while in Iowa City that summer.

> He took great pride in telling me about his summer at the University of Iowa and seeing my name on the grade sheets for the organic chemistry course I had taken the previous spring. In those days, mid-term and final grades for each course were posted for all to see. In case you are curious, I made an 'A' in organic chemistry, mainly because a fellow student called me a 'curve raiser,' which he did not feel was fair to the pre-meds. After all, I was planning on being just a teacher! (N. D. Anderson, personal communication, September 12, 2014)

Norm graduated from high school on May 23, 1946, and was awarded the Bausch & Lomb Honorary Science Award. He cites this award as an additional stimulus to study science in college. Following Norm's graduation from high school, he enlisted in the U.S. Army and began his service on June 12, 1946.

After a long, stormy voyage to Yokohama, Japan, in February of 1947, Norm was assigned to the 738th Engineering Battalion and spent his working hours supervising the loading of lumber onto cargo ships in the

Yokohama harbor. His stay in Japan would be punctuated by many outings around the countryside and including a hike to the treeline of Mt. Fuji. Fog the following day prevented the final ascent to see the volcanic crater. In October of 1947, Norm had a return voyage to the United States.

After receiving an honorable discharge from the U.S. Army, Norm resumed his plan to get a college degree. Initially, he studied at a small college in Sioux City, Iowa, before transferring to the University of Iowa with plans to become a science teacher. Although Iowa State University was much closer to the family farm and was a preferred college destination among his peers, he chose to go to Iowa City and get a degree in general science. Fast forward a couple of years and Norm was on track to finish the requirements for a BA degree one semester early, but had been unable to find a job. However, he soon learned of a teaching position in a junior high school in Burlington, Iowa, which was to become available at the midpoint of the school year. He interviewed, was hired, and soon finished his exams for the fall semester of 1951. To start this new career, he moved to Burlington in a New Year's snowstorm. The principal of the junior high school in Burlington had offered Norm a place to live while he settled in. The principal told Norm that he always went to school very early but for him to come on down to the school as soon as he was able the next morning. He arrived at the school about 7:30 a.m. and was assigned to teach in an upstairs room of the building. Across the hall was another teacher (who, unbeknownst to Norm, was to become his brother-in-law) and the two of them sorted the students who were sent to them into the two science classes per period. His instructions were to simply ask the student what grade he/she was in and then to teach them. At the end of the day, he had four classes full of students, but the full load for a teacher was five classes in a six-period day. So the principal asked if he could teach a 7th-grade geography class. Norm did not feel qualified to teach geography to any age group, but accepted the challenge and was greeted by 39 lively and very bright students. For an initial assignment, he asked the students to write an article about themselves. He reported that he was amazed by what he learned about the students and their families; this knowledge was to prove of great value to him as their teacher, but also served as a model to use later as a teacher educator who was to encourage prospective science teachers to find ways to know their students and thus know more about how to teach them. Inquiring what the first topic for the second semester of the course should be, he was told that the previous teacher had covered Europe and other geographic sites, but to this point there had been no mention of South America. Norm soon discovered that there were 13 countries in South America and that the total of thirty-nine students was divisible by 13 to create groups of three students per group. Thus, as much due to the serendipity of the mathematics as for any other cause, he decided to teach the geography of South America and created a project-based course, for which each group of students took one country in South America, and did a group project to share with their peers. The fact that the physical size of the countries is widely variable (e.g., Brazil was much larger than the others) was deemed irrelevant. The students used the library, available copies of *National Geographic* magazine, and any other resources that could be found to create their projects. But Norm learned a second important lesson for his future teacher education career about getting students busy and letting them do much of the work. Talking of this experience many years after its occurrence, he described the insight of realizing that "if you can find a strategy or approach that lets the students do some of the work, then it will work best for you too."

Norm began working on a master's degree while he was teaching and in 1956 received this second degree. At this same time, science education as a discipline was beginning to be fueled by very large amounts of money from the federal government. In 1950, the National Science Foundation had been formed by the U.S. Congress; the federal agency would not start specifically supporting the U.S. science teaching mission until the mid-1950s. But in addition to the curriculum development efforts that would be sponsored by the NSF from the late 1950s into the 1970s, two types of institutes were created to support science teachers in raising their knowledge of science and science teaching. The first was called "summer institutes" and were usually hosted by university science departments. Science teachers would apply for admission and were typically paid an honorarium for attending a 4- to 8-week summer experience. The second type of institute was called the "Academic Year Institute" and hosted in-service teachers at a college or university for the entire academic year. It was not uncommon for academic year institute attendees to acquire or at least make progress toward acquiring graduate degrees. In Norm Anderson's case, an academic year institute starting in the fall of 1957 was the first step in his becoming a doctoral student in science education at The Ohio State University. The fact that Norm had arrived for the academic year institute in the fall of 1957 also serves as evidence to invalidate the commonly held belief that the all science education reforms of the 1950s and 1960s were initiated as a response to the Soviet Union's launch of the Sputnik satellite. Norm began his academic year institute in the fall semester of 1957 and on October 4 of that same autumn Sputnik was launched.

As stated earlier, Norm was to work his way through this institute and ultimately his entire doctoral program under the tutelage of Dr. John S. Richardson. At almost the same moment as the matriculation into this second round of graduate studies, Dr. Richardson had published a secondary science instructional methods book titled *Science Teaching in Secondary Schools* (1957). The introduction to that book gives an encapsulation of Dr. Richardson's thinking about the important issues in science education of that day—and the emphases that were encountered by Norm as a student. This introduction will be quoted at length:

> Our society is in the process of developing scientific ways of thinking and acting; it is enjoying the material benefits of the applications of science to our problems. Progress in developing critical thinking is seemingly slow, but the development of material benefits is re-emphasized each day with new products and applications of our technology. As ignorance, prejudice, and superstition encounter enlightenment and critical thought, conflicts occur; as the products of advancing technology change our patterns of living, other conflicts occur. The resolution of such conflicts is brought about most satisfactorily through planned educational processes.
>
> Those who have competence in science and responsibility for science in the school curriculum face problems that transcend the transmittal of content. They must be concerned with the social and economic implications of science and technology. They must carry the responsibility for the advancement of science in all its directions through their teaching. Most critical of all, they must stimulate and nurture the creative intellect.
>
> Those who are or will be science teachers or are otherwise concerned with science in our schools face such a pyramid of responsibility. To them this volume is directed. (Richardson, 1957, p. v)

John Richardson had been trained through his bachelor's and master's degrees as a chemist, but in the interim before his doctoral work, he had served as teacher and principal of the McGuffey School in Oxford, Ohio, and as a member of the faculty at Miami University of Ohio. Through these experiences, he moved into science education as the profession through which he would direct his intense focus. Richardson would return to Ohio State to complete his doctorate and be retained as a faculty member in science education. Norm reflected on how the College of Education at Ohio State, and Richardson in particular, were champions of the Progressive Education movement led by John Dewey, though this movement had reached its zenith just prior to WWII. As Norm was arriving at Ohio State, the science education program was moving into a facility new to them, which was in fact a part of the old library building. Norm was given the job of organizing and arranging teaching-related equipment and other materials in the new space while also cleaning out what had been left behind. Much of the science equipment was army surplus, available in great quantities in those postwar years. But one of his most important discoveries among the trash and treasures accumulated in that space was the handwritten manuscript of a speech given by John Dewey at The Ohio State University some years earlier.

When Norm arrived at Ohio State in the fall of 1957 to begin the academic year institute and his doctoral work, he was accompanied by his young family—one that would ultimately include six children. One weekend, he took his children on a "walk in the woods" and discovered an old log that was rotting and was covered by a great variety of decomposed organisms. He had been assigned to teach a general science teaching methods course and so he brought this rotting log to class and posed the question to his methods students: "How many ways can you come up with to use this log in teaching science?" Another of the science education professors, Dr. Fred Schlessinger, was scheduled to use the classroom during the following period and learned about Norm's use of the rotting log. Soon, this activity became known to Dr. Richardson, and he found it to be perfectly in keeping with his ideas of the student-centered classroom at the heart of the Deweyian philosophy of teaching. After that fortuitous event (at least in the eyes of Dr. Richardson), Norm was acknowledged as having a serious understanding of the student-centered classroom. This event would even carry over into the time when Norm was in the job market seeking a new science education professor position at NCSU. In his letter of recommendation to support Norm's application for the position, Dr. Richardson described Norm's teaching ability. He wrote,

> Frankly, I have been amazed with the quality of his teaching. Naturally, I had expected that he would do a good job, but I was not prepared for the kind of performance he would turn in. I think I have never seen better teaching in this course. His enthusiasm and imagination bring out in the finest way, his fundamental competency in the field. (Speece, n.d., p. 5)

Prior to receiving his PhD, then as now, a dissertation had to be put forward and defended. And then as now, that dissertation had to be judged by a committee as to whether it would make an original and significant contribution to the research and scholarship of its field. Norm's dissertation was titled, *An Analysis of Programs for the Preparation of Secondary-School Science Teachers.* The problem investigated in the dissertation was described in the Abstract as follows:

> This investigation resulted from a desire to know how science teachers are presently being prepared for secondary-school teaching. The problem investigated in the study was stated as two questions. (1) What are the criteria that can be used in analyzing selected elements of the academic and professional phases of programs for the preparation of science teachers? (2) Using these criteria as a basis of judgment, what is the status of programs in five states—Illinois, Indiana, Michigan, Ohio, and West Virginia? (Anderson, 1965, p. 1)

Using the research literature of science education, Norm had created a set of nine criteria to reflect the teacher preparation program emphases put forward for secondary science teachers. He submitted the criteria and related supporting statements to a "jury of ten science educators." From this validated set of criteria, a questionnaire was created and sent to 155 institutions that were preparing science teachers in those five states. Norm obtained a response rate of about 50%, with approximately two thirds of the institutions that responded being private colleges and universities. Most of the responding institutions prepared fewer than 10 science teachers during the 1962–1963 school year, and about half of the science teachers prepared were specialists in biology. These statistics suggest that little has changed in the 60 years since Norm completed his dissertation, at least in terms of the size of programs for science teacher preparation and the science content emphases.

It is instructive to our present-day efforts to consider the criteria that Norm found within the literature of science education regarding the important or significant characteristics of secondary science teacher education programs. They were

> (1) cooperative efforts of scientists and professional educators; (2) provisions for general education, professional education, and academic preparation in the sciences; (3) breadth and depth of preparation in sciences; (4) special science courses and experiences for science teachers; (5) development of professional competencies; (6) professional laboratory experiences; (7) special facilities and resources; (8) qualifications of selected faculty members; and (9) experimentation. (Anderson, 1965, p. 1)

It seems likely that each of these would be among the set of criteria developed from the literature of contemporary science education for a similar study conducted now.

Findings from his dissertation study were also quite relevant to our science teacher preparation efforts in the second decade of the 21st century. Although it was recommended at that time for 50% of the coursework taken by prospective teacher to be in the sciences, "a large majority" of the institutions did not have this requirement in place for their prospective science teacher education students. Many of the institutions also did not have subject-matter-specific science methods or curriculum courses available for the students who were prospective science teachers. Faculty members in the sciences were the most common instructors for the courses dealing with the teaching of a specific science subject matter. During student teaching it was relatively rare for a student teacher to be supervised by a faculty member with a science specialty.

Norm arrived at NCSU as a new assistant professor for the fall semester of 1963, before finishing his PhD degree but with a great load of data. Almost immediately he realized that the science educators at NCSU, Dr. Herb Speece and Professor Henry Shannon, had already modified the secondary science teacher education program to meet many of the recommendations that would eventually arise from his dissertation research. Norm put himself to the task of writing the dissertation and also to continue the books related to science education that he would write.

His first book was already published before arriving in Raleigh for his position at NCSU. This book, titled *Challenge of the Universe,* was co-authored with Allen Hynek, who had become famous for his research into the existence of alien visitors from other worlds, or to use the term that had not long before been coined by the U.S. Air Force, UFOs. In the article commemorating his designation as an Alumni Distinguished professor, he said this in response to questions about his work with Hynek on that book.

> When asked about his own belief in unidentified flying objects, Anderson explained his "three box" system for dividing phenomena. "The first box is labeled definitely 'Science.' The second box says 'Philosophy and Religion'. And the third box has a question mark on it—it's the biggest of the three. UFO's, ESP parapsychology, superstitions like planting by the moon all go in this one. There is enough evidence to warrant serious investigation of these, but not enough to label them 'definitely science.'" (Profiles, 1978)

Through the late 1950s and into the 1960s, science education was undergoing a transfiguration under the influence of huge financial support from the National Science Foundation, but also from the public hysteria resulting from Sputnik and the general frenzy caused by the Cold War. A large portion of the reform that was underway came in the form of new curricula for the science subject areas. This effort was largely led by scientists such as Dr. Jerrold Zacharias, a physicist at the Massachusetts Institute of Technology (MIT), who was compelling forward the creation of the *Physical Sciences Study Curriculum* (PSSC). In response to a question, Norm related that he and Dr. Richardson shared concerns about the dominant role of the scientists. Dr. Richardson, who was an intense professor, had concerns such that a graduate student needed to exercise caution when discussing these projects in his hearing. One of those concerns regarded the role of the scientists who were seen as creating curriculum materials without adequate appreciation of the role of and potential contribution from science educators or teachers. Perhaps Richardson had felt the sting of the put down of the science education community reported by Rudolph (2002) in his book, *Scientists in the Classroom.* Rudolph described the situation this way:

> Not long after the announcement of the formation of the National Committee, an official from the National Association for Research in Science Teaching (NARST) contacted the Committee's executive secretary, Clark, lobbying hard for the inclusion of his organization in the mix. There should be "closer cooperation between scientists and science educators," he argued, in working "to solve the shortage of scientists and science teachers." Clark, unfamiliar with the organization, ran the request by NSF staff members for

> guidance before taking further action. No ringing endorsement was forthcoming. One Foundation official counseled that the "group is OK, but not outstanding." (p. 80)

Norm related that during this period of time, it was often difficult to express a differing or dissenting opinion with regard to one of these curriculum projects in a public or professional forum without drawing an acrimonious response from one of the curriculum's authors or stakeholders. He related a memory of an article in a professional journal where a science teacher wrote a description of how he felt that the curriculum units worked better in his own classroom if they were arranged somewhat differently than in the new curriculum plan. Even this minor adjustment drew ire. But Norm was also quick to say that it is important to try new things, because that is the only real route to change. It is also the basis for much of the actions that have driven science and technology forward in the past.

One incongruous aspect of the so-called alphabet curricula that Norm identified was the diminished emphasis within science books and curricula on technological applications used within our society that had resulted from science. Examples of such technology were, for instance, the refining of raw materials that were found in mineral deposits or even the technology used to remove these raw materials from the ground. As he pointed out, one goal of the new curricula was to put emphasis on "new content" that science had produced in the modern era, and thus some of those older technologies were seen as anachronistic. But there is a certain irony to now looking at the emphasis on STEM in our current educational environment and realizing that some of those older technologies would be quite useful to teach the integration of science-technology-engineering-mathematics. Most of the technologies that were used in the alphabet curriculum materials were put there to create an emphasis on the scientific laboratory.

In 1968, NCSU Provost Harry Kelly (who had been a science adviser to President Harry Truman and had been at senior administrator at the NSF prior to coming to NCSU) worked with the Alumni Foundation to create a Distinguished Professorship designation to be awarded to those members of the faculty who were identified as particularly notable classroom teachers. Norm Anderson was selected to receive this designation, which had been formally named "Alumni Distinguished Professor" in 1973. He was one of only 23 faculty members to receive this recognition in the first 10 years of the award program.

When asked about his philosophy of teacher education, Norm did not point to any one aspect of his preparation or experience as a guiding factor. Rather, he said,

> I don't know that anyone ever knows that stuff—part of my philosophy working with [instructional] methods is that it is more coaching than telling. The

> methods instructor is not imparting knowledge in the same way as a professor of physics or chemistry does—and the students come with all kinds of experiences in school—at least 12 or 13 years—you are trying to get them to self-examine what they are doing [when learning to teach]. (N. D. Anderson, personal communication, November 20, 2014)

He went on to say that in the years of his own teacher preparation and even into his doctoral work, the prospective science teachers did not "get much help."

When Norm first arrived at NCSU, the students took a sophomore-level Introduction to Mathematics and Science Education course. As the semester progressed, the program "slowly worked into sending the students to schools as teachers' aids." Just as is often true today, the major route by which undergraduate students came to the secondary science teacher education program was as transfer students. Students change their minds about the allure of teaching as a career and so it is not a matter of getting students who cannot find an academic home in the university, it is a matter of the student's maturity helping to refocus his/her career goals. Often in this refocusing, teaching begins to appear as a more satisfying career option. Research dating back to at least the 1950s has demonstrated this fact (e.g., see Langston, 1951). Much of the classroom portion of the secondary science methods classes that Norm taught were aimed at developing skills to help the prospective teachers conduct safe and effective laboratories and demonstrations. As the program began to send the prospective teachers to schools for a greater portion of their coursework time, there was a corresponding reduction in the time left for teaching the laboratory aspects of *teaching* science. But as always, Norm reflected on how the time spent in schools allowed the prospective science teachers an opportunity to acquire models of pedagogy and the other aspects of classroom life.

The most powerful role model of Norm's professional career was Dr. John S. Richardson. Richardson was a prolific scholar who published a great variety of books and articles regarding science teaching and learning. Among Dr. Richardson's many accomplishments was being one of the co-founders of the journal, *Theory into Practice*. Richardson was a strong believer that one's work, whether in the classroom or the profession, should be directed at producing products that were disseminated and used. In his seminar classes, he would have the doctoral students write focused manuscripts on important topics of that day and then collect and publish them in book form. Unfortunately, John S. Richardson did not live to see Norm promoted to the rank of Full Professor. He died on May 22, 1969, at the age of 60.

Norm felt that he had some great professors in both science and education coursework. Other educators and scholars who were powerful influences on Norm included Dr. Marge Gardner, a professor at the University of Maryland and a powerful influence on science education in the 1960s through the 1990s; Dr. Walt Brown, a frequent collaborator and co-author of many books; Dr. Jack Shrum of the University of Georgia, on whom he often relied for advice on how to handle many problems in science education; Norm's first principal at the junior high school in Burlington, Iowa;and teachers and others in his high school in Spencer, Iowa. He also feels deeply indebted to the mentoring and colleagueship he received, starting from his arrival in Raleigh as a new professor at NCSU, from Dr. Herb Speece, who was the first mathematics and science education professor at NCSU, and from Professor Henry Shannon. One of the most important professional attributes that shaped his career was Norm's emphasis on teaching junior high school science. (For those readers who might not be familiar with junior high schools due to the prevalence of the "middle school" model, especially in the United States, the junior high school was typically the schooling site for grades 7, 8, and 9. Frequently, teachers were divided into academic subject-matter-based departments in a manner similar to high schools. However, the grade-level arrangement was not considered convenient, as the 9th grade course credits were counted toward Carnegie Unit totals required to receive the high school diploma. Also, in the middle school, the academic subject-matter departments were replaced by interdisciplinary teams of teachers who taught a common set of children. Today, middle schools outnumber junior high schools by a ratio of 10 to 1 according to Wikipedia [http://en.wikipedia.org/wiki/Middle_school].) As was mentioned earlier, Norm became a junior high school teacher because of the availability of a job in Burlington, Iowa, but he had been interested in teaching at that level from the very beginning, and this interest was one reason he chose to major in general science rather than a more specialized science field. However, as a science education professor, he discovered that there were few other science education professors around the United States who had either taught at the junior high level or were emphasizing those grade levels in scholarship. Rather, most drew on experiences at the elementary or high school levels, and as a result, reform in junior high science was somewhat less emphasized than at the other levels. Thus, he became a specialist in this area. His science textbook series (written with his frequent co-author Dr. Walt Brown) had two editions (the first published in 1972 and the revision in 1977) for the three grades of the junior high school. In the textbook series, and generally throughout the nation at that time, the seventh grade was home to the life science course, followed by earth science in the eighth grade and then physical science in the ninth grade.

I once read a book titled *The Book of Prefaces* by Alasdair Gray (2000), in which an anthology of prefaces of important books published between the 7th and 20th centuries told a story of various authors' motivations to write their books and reflected our civilization's ascent. Gray gave credit to an 18th century author named William Smellie for describing the motivation that led him to create this edited book. Smellie had written, in the preface to his book, *The Philosophy of Natural History*, which had been published in Edinburgh in 1790, that

> Every preface, besides occasional and explanatory remarks, should contain not only the general design of the work, but the motives and circumstances which led the author to write on that particular subject. If this plan had been universally observed, a collection of prefaces would have exhibited a short, but curious and useful history both of literature and authors. (Gray, 2000, p. 7)

Earlier in the chapter, I used a portion of the introduction to Richardson's science methods book to set a context for the doctoral program into which Norm entered in 1957. But now, drawing on this same source of motivation, I have chosen to pull together the loose ends of this story using the introductions to two books that Norm wrote during his long and productive career. Both books were written with co-authors and both were directly aimed at science teaching. The first is the first edition of the junior high physical science book that he co-authored with Dr. Walt Brown titled, *Physical Science: A Search for Understanding* (1972) published by J. B. Lippincott & Company. The second is the secondary science methods book co-authored with Dr. Ronald D. Simpson titled *Science, Students, and Schools: A Guide to Middle and Secondary School Teacher* (1981). Neither of these introductions can be quoted at great length, but looking seriously at what he and his co-author wrote, when they had time to deeply consider what they truly believed about science teaching and learning, is a better gauge of the man than perhaps any other source.

Very early in the introduction to the teacher's edition of *Physical Science: A Search for Understanding*, the reader encounters a section labeled "Why Do We Teach Science?" Their answer to this most important question is as follows:

> To answer this question, we pose two additional questions. What is the nature of the society in which our students live? What is the nature of the scientific enterprise?
>
> Our society is first a democratic one, placing emphasis on the optimum development of each individual as a contributing citizen. In addition, it is a scientific and technological society, in which new career opportunities are constantly being generated. Also, ours is a rapidly changing society, not only scientifically and technologically, but in all areas of human endeavor. The result is a need for future citizens who are able to solve a multitude of problems using rational scientific processes.
>
> The current concept of science is that of a human activity—one of many techniques civilized man has developed to aid him in his search for "Truth." This is quite different from the definition of science as "a collection of facts," popular only a few years ago. In light of this new philosophy, a list of purposes

> of science teaching must include a great deal more than the transmission of the *products* of science—the theories, hypotheses, and concepts currently accepted by the scientific community. Additional objectives must include the concept of science as a structured, logical process: understanding of the tentative nature of scientific theory, and thus an understanding of both the self-correcting nature of the scientific enterprise and the cumulative nature of scientific knowledge; and the development of skills of objectivity, logical thinking, and tolerance on the part of the student.
>
> Thus, science should be taught in such a way as to help produce a scientifically literate citizenry—a citizenry that has knowledge of the *products of science,* of the effects of these products on his society, and of the relationships between science and technology. A scientifically literate citizen also will have some understanding of the *processes of science* and their implications for the solution of many types of problems—not just those of the laboratory. (Brown & Anderson, 1972, p. T-6)

Norm and his co-author go on to write in detail of what it means to consider the products of science. Included in this description is the report of a panel assembled in 1963 by the National Science Teachers Association (NSTA) to identifying the conceptual schemes and major items in the process of science.

Additional sections in the introduction include directives on approaches to teaching and means to help students learn. They put particular emphasis on student learning, but also generalized this to learning by people in general. They wrote,

> In a rapidly changing environment, one need of all people seems apparent. This is the need to be able to learn effectively. If we cannot predict the type of occupation a particular student will have, or the type of leisure-time activities he will have available to him, or even the type of family relationships he will establish, we can predict with certainty that he will be constantly required to make adjustments to changes in his life style. To do this efficiently, he must know how to learn. (Brown & Anderson, 1972, p. T-10)

In the preface to the 1981 edition of *Science, Students and Schools,* Norm and Ron structured their thoughts around a set of guiding principles which "if this discussion were given a title, it would be 'In These Things We Believe'" (Simpson & Anderson, 1981, p. vii). Norm had used this phrase earlier when he was interviewed for the *Profiles* document created to commemorate the Alumni Distinguished Professors at NCSU in 1978. In the article for that volume, written by Ruth Yachan, Norm was quoted.

> "Talking about teaching 'techniques' doesn't turn me on," Anderson said. He believes more in using whatever method works best in each specific situation. "There's no one way. I like using a variety of approaches."

> However, Anderson has some basic standards which he practices as well as preaches. He has a file called, "In these things I believe," which contains quotations that he finds helpful.

Thus, the introduction to this methods book as well as his basic philosophy can be found to have roots dating back to Mr. Myhr's science class at Spencer High School so many years earlier.

In describing those things in which they believed, Norm and Ron started with the belief that "science and technology have played a major role in the movement of humankind from a primitive, harsh environment to a world of knowledge, understanding, and increased comfort" (Simpson & Anderson, 1981, p. vii). They believed that this change in society also was responsible for the creation of a degree of freedom, ability to make choices in society, but also responsibility to support the education of the citizenry to be scientifically and technologically literate. The list continued with the statement that the "young people are our country's most precious resource and that they come in all shapes, sizes, and temperaments" (Simpson & Anderson, 1981, p. vii). But they also believed that the nature of our society is inextricably linked to science and that science education must be regarded as very important within the school's curriculum. There was also a tip of the hat to the importance of recognizing that there is no one way to teach science regardless of how the task may have been conceptualized by curriculum developers or scientists. They wrote,

> Finally, we are aware that there are many ways to organize and conduct science instruction and to evaluate students. The mark of professional science teachers is the repertoire of skills they use in the classroom and the ways they interact with students. Effective teaching embodies many skills that can be learned and improved upon over time. We have drawn from many years of our own experience, from the experience of others, and from a broad range of literature and research in an effort to describe the state of the art of science teaching and to offer practical suggestions for ways science teachers can continue to improve instruction. (Simpson & Anderson, 1981, p. viii)

When asked what projects he would have liked to have done, had time and resources allowed, he said,

> My goal was to create a dictionary of science education. I created a list one time of terms of the day such as formative evaluation, summative evaluation, and so forth. Education has always been accused of having a lot of jargon—but you know what the definition of jargon is don't you... it is "the technical language of someone else's field." The dictionary might have included the two best references that illustrate the context in which that term was used in that field. (N. D. Anderson, personal communication, November 20, 2014)

Asked to describe his greatest contribution to the field, Norm was reluctant to point to any product or specific activity. He took pride in the fact that he had published all of those books, though he wished he had been able to write more. He took pride in the fact that he led the successful effort to create a PhD program in mathematics and science education at NCSU soon after he arrived. More than anything, he reflected on his intersections with people. He felt great satisfaction from the knowledge that many of those individuals, students, faculty, and others, who met with him felt that they had received good advice and mentoring. As he knew then, and many of us know now, at large universities, "many kids fall through the cracks" but our ability to be there for them "is what it is all about." Being an exemplary citizen of the university and of the profession were roles that were important for him to play. Many of us needed that advice and mentoring. He relished his ability to play those roles. As part of his successful nomination to be included in the College of Education Hall of Fame at The Ohio State University, Dr. Ron Simpson and Dr. John Shrum wrote the following about Norm: "Here is an uncommon man who devoted his life to helping others, and in doing so, influenced a generation of science teachers and leaders in elementary, secondary and higher education" (Shrum & Simpson, 2004, p. 1).

And so we come to the conclusion of this brief biography of Norman D. Anderson. I suppose that it will be incomplete without my own statement of why I sought the opportunity to assemble this record of his accomplishments. I was one of those students at NCSU who decided that science teaching was a career in which I could succeed and so in 1976 decided that I needed to get degrees in both zoology and science education on my way to becoming a science teacher. Norm Anderson was one of the professors who attempted to shepherd me through that second degree. But more importantly, some years later, when I had been a science teacher for 4 years and needed to find a mentor and a direction, I returned to NCSU to work on a master's degree. I told Norm very near the beginning that I wanted to learn to do educational research. As a young teacher, I had heard people talk about this activity, but I really did not know what it meant or whether I would have any propensity to do it. Norm again took the role of shepherding my degree program and gently leading me through a study of more than 400 eighth-grade students regarding their attitudes toward science teaching as a career. The completion of this study resulted in my receiving an MS degree from NCSU, publishing one of my first articles in a major journal (see Oliver & Anderson, 1986), setting a course to pursue a doctoral degree, and ultimately becoming a professor myself. The time spent in that master's degree program and the direction that it gave to my professional aspirations was the most important career-related event of my life. And to commemorate that, in 2007, my late wife Jenny and I created a scholarship at NCSU to honor Norm Anderson and to aid a few of those

Figure 7.1 Dr. Norman D. Anderson

students who will follow, in hopes that they too will find the mentoring and direction needed to accomplish what they have the potential to do.

At the time of this writing, Norm Anderson lives in Raleigh, North Carolina, and fills his time with family (the six children have produced many grandchildren and now great-grandchildren), woodworking (his specialty is making wooden Christmas tree ornaments of birds), and writing his newsletter called *The Ferris Wheel News,* which recently had its 179th issue. Many of the stories and personal reflections included in this chapter were related to the author during telephone conversations that happened on September 12, 2014 and November 20, 2014.

REFERENCES

Anderson, N. D. (1965). *An analysis of programs for the preparation of secondary-school science teachers.* (Unpublished doctoral dissertation). Columbus: The Ohio State University Press.

Anderson, N. D. (1986–1987, Winter). I remember Springdale School. *The Rural Educator,* 1–3.

Brown, W. R., & Anderson, N. D. (1972). *Physical science: A search for understanding.* (Annotated ed.). Philadelphia, PA: Lippincott.

Gray, A. (2000). *The book of prefaces.* London, UK: Bloomsbury.

Langston, R. G. (1951). A study of attitudes toward teaching as a vocation. *Journal of Teacher Education, 2,* 83–87.

Oliver, J. S., & Anderson, N. D. (1986). Science teaching as a career choice of junior high school students. *Science Education, 70*(2), 93–103.

Profiles. (1978). *Profiles: North Carolina State University Alumni Distinguished Professors.* Raleigh, NC: NCSU.

Richardson, J. S. (1957). *Science teaching in secondary schools.* Englewood Cliffs, NJ: Prentice-Hall.

Rudolph, J. L. (2002). *Scientists in the classroom: The Cold War reconstruction of American.* New York, NY: Palgrave Macmillian.

Shrum, J. W., & Simpson, R. D. (2004, June 23). Letter to Hall of Fame Selection Committee at the College of Education Alumni Society at the Ohio State University.

Simpson, R. D., & Anderson, N. D. (1981). *Science, students, and schools: A guide for the middle and secondary school teacher.* New York, NY: Wiley.

Speece, H. E. (n.d.). *A short history of the Department of Mathematics and Science Education at North Carolina State University.* Unpublished manuscript.

Yachan, R. (1978). Norman D. Anderson. In *Profiles: North Carolina State University alumni distinguished professors.* Raleigh, NC: NCSU.

CHAPTER 8

HANS O. ANDERSEN

A Lifelong Mentor and True Friend to Teachers

Teddie Phillipson-Mower
Indiana University–Bloomington

Hans Oliver Andersen (1935–2014) was a beloved science educator, mentor, and friend to teachers. He spent his professional life dedicated to improving science teaching at all levels and advocating for the science teacher. Andersen's illustrious career spanned 45 years and made a positive and lasting impact felt in Indiana, nationally, and around the world.

THE EARLY YEARS

Hans Andersen was born on March 30, 1935, in Wisconsin Dells, Wisconsin, a perfect environment for exploring the natural world and setting a course for a life of advocacy and service in science and environmental teaching and learning.

His parents, Henry and Esther Andersen, were the owners of a shoe store where they sold and repaired shoes. The family of six lived on the second

Going Back for Our Future II, pages 129–151

floor of the store. Patricia Buob, Hans' youngest sister, described the family as "not having very much, but rich in so many other ways" (P. Buob, personal communication, October 15, 2014). She recalled the family garden that hugged the sides of the house to yield space for customers to park their cars in the prime locations of the big backyard. They were a family of gardeners. Their father harvested seeds from flowers and vegetables he found as he meandered the neighborhood, which gave rise to a wonderfully diverse garden which represented almost every plant grown in the area. Hans and his siblings also enjoyed roaming the countryside foraging for blackberries and other edibles, traversing property lines, and staying out for hours on end.

The family spent a lot of time outside, hiking and exploring local natural areas. There was a lot to discover in the Wisconsin Dells. Ross' Ravine, Rocky Arbor, Witches Gulch, and Louis' Bluff are just a few of the prominent landforms that captured the imagination of young Hans. Henry took many pictures of the family as they investigated the natural world together, often taking them out later to quiz the children about the names and features depicted in these documents of their family life.

Hans loved sports, especially track and football. Wisconsin Dells High School was a very small school and he joked that it was really easy to excel when you are the only one on the team. Hans and other young people in the community would gather at every opportunity to play football. Patricia reminisced about playing football with the other kids. She was one of the youngest players but gave it her best. She loved playing with her big brother. Hans allowed her to play and was very encouraging. He was also very protective. That went both ways, she shared. "When the other kids got mad at Hans, I would yell at them" (P. Buob, personal communication, October 15, 2014).

Patricia said that because they went to a small school there were few options for classes and programs of study. While Hans was immensely curious and loved learning about and in the natural world, there was little evidence of an academic interest in science in Hans' early years. Their mother had been a teacher but could not continue her education due to sickness and depression. Therefore, the Andersen children were never allowed to say, "If I go to college." It was always, "When I go." For young Hans, the opportunity and passion for higher education didn't come until after serving in the Army (P. Buob, personal communication, October 15, 2015).

Hans met Sandy, his partner and editor for life, at the University of Wisconsin. He had enrolled in business as a freshman, which did not excite him. It was mutual. Sandy said, "[the Business School] actually encouraged him not to come back" (S. Andersen, personal communication, August 23, 2014). He decided that he wanted to go overseas to Korea and joined the Army as a cook. One month later, the war was over. After he was released from the Army, Hans reenrolled in the University of Wisconsin–Whitewater and took his first biology class, with Henry Meyer. He couldn't get enough.

Professor Meyer, who received an Excellence in Teaching Award from UWW, took Hans under his wing. Patricia said, "Once in school, he devoured it." (P. Buob, personal communication, October 15, 2014). He took every science class offered at the Whitewater campus. He graduated with his undergraduate degree in 1959.

After Hans attended a summer biological institute, he decided to continue in biology and enrolled at the University of Michigan. Hans received his MS in biology in 1960. He considered pursuing a PhD in comparative anatomy to prepare for a career of teaching at a medical school and was highly encouraged by his advisors to do so. He knew he wanted to teach. The question was, in what context (S. Andersen, personal communication, August, 21, 2014).

Hans and Sandy were starting a family and he needed to figure out what he wanted to do. When a friend offered his summer house in the Dells as a place for contemplation, they gratefully accepted. Hans took on a variety of jobs during this time. He later told his son, Scott, about how much he enjoyed working as a guide on the Wisconsin River and playing Smokey the Bear for the tourists. He loved meeting people and sharing stories of the natural history and science of the environment he had explored as a child (Andersen, 2014). Hans also worked at a gas station and pea factory to make ends meet. As the summer progressed, he decided that he would like to teach high school biology. A former colleague of his was working for the Niles Township Community Schools and told him that there was a science position opening. Hans accepted the position and the family moved to Skokie, Illinois (S. Andersen, personal communication, August 21, 2014; P. Buob, personal communication, October 15, 2014).

It was the right move. He put his heart and soul into teaching. Within 2 years he became the science department chair, a position that gave Hans an opportunity to work with teachers in all stages in their careers. He became curious about how to best support quality science teaching and advocate for the teaching professionals whom he worked with both directly and indirectly. In 1962, Hans and teaching colleague Jim Winkleman explored how team-teaching methods might enrich a one year biology course. This resulted in Hans' first publication, "An Approach to Team Teaching in Biology" (Andersen & Winkleman, 1962). Always seeking additional learning opportunities, Hans participated in an NSF Institute in geology that was held at the Indiana University (IU) campus in Bloomington in 1964. Sandy said, "He came home from the Institute one day and asked, 'Do you think you could live here? I want to start school in September'" (S. Andersen, personal communication, August 21, 2014). They had to wait until January 1965 to make the move, but they put their house in Skokie on the market, moved to an apartment in the old Army barracks in Hoosier Courts in Bloomington, and Hans enrolled in the IU EdD program. He completed his EdD in 1966, passing his dissertation defense, "An Analysis of a Method for Improving Problem-Solving Skills Possessed by College Students Preparing to Pursue Science Teaching as a Profession" (H. Andersen, personal communication, March 16, 2014).

Hans took a position as an assistant professor of science education at IU in 1966. Even though Hans looked into other positions over the years, both he and Sandy loved Bloomington, and the decision was always to stay. He thought of his students as colleagues, working together to improve science education teaching and learning. He encouraged them to address him as Hans instead of professor or doctor, terms that to him were used to create a divide that often put teachers and teaching at a lower level instead of one of priority. In that first year Hans and Sandy also started the tradition of hosting science education gatherings where science teachers, students, and faculty from IU and other universities would come together to forge relationships and share ideas. He was there to serve, nurture, and advocate for his students and for science education.

AJARN (HONORED TEACHER) HANS

The Triad

Dr. Hans Andersen's first task as assistant professor at IU was to develop the secondary science methods course, which framed his early scholarship in the areas of curriculum development, problem solving, and teaching strategies. As part of an early cohort of science educators who championed the articulation of instructional objectives in the form of behavioral terms, he organized the science methods course around their use in order to provide direction for the type of instruction that was necessary for the future science teacher. In a well-cited article in the *American Biology Teacher*, "Science Methods Instruction" (1969), he laid out a list of attainable objectives that should be included in both the theoretical and laboratory sections of the preservice science methods course. He argued that preservice training should equip students with tools to use in the current teaching environment as well as prepare them for the future that will bring "mechanisms of change and reform." His publications at this time and his willingness to share concrete, practical applications drawn from theory led to a sweeping change in the way science methods courses were structured and taught. In addition, national and international colleges and universities invited Hans to work with them in improving their science education program and courses.

Terence (Ted) Mills, retired professor of science education from Oklahoma State University, called this Hans' Mager phase, in reference to Robert Mager, who first made the case for the use of behavioral objectives instead of "generalities" to improve teaching. In a letter that Ted Mills wrote in congratulating Hans on his retirement, he said,

> I will never forget the "task analysis" of a four letter word you assigned to John Rusch, Steve Godomsky and I. We figured this was a piece of cake and our four letter word "heat" would present no problem. Weeks later, the results of our analysis produced an entire wall covered with sequenced labels for the prerequisite facts, concepts, and principles for "heat" yet we were still no closer to an answer to the full meaning of the concept! As a result of this experience I believe I learned more physics than I ever had in any single science course! Thanks for that!! (T. Mills, personal communication, September 10, 2002)

John Staver, professor of curriculum and instruction at Purdue University, reminisced that research at that time, at least in Hans' eyes, never looked at the triad—research, teaching, and service—as separate entities; he always saw them as connected. John stated that he shared that perspective. "I wanted to know more about how to do it, what good teaching is, what bad teaching is and all of the issues related to it" (J. Staver, personal communication, April 19, 2014). John's thoughts are demonstrated in the example in the preceding paragraph. For Hans, research, teaching, and service should always be a seamless interconnected triad with the goal to improve science education.

As his career advanced, Hans continued to hold strong and advance his commitment to science teachers and science education training in his scholarly activity. He didn't have a lot of patience for jargon or theory that wasn't grounded in practicality, since he felt that it led to a hierarchical divide that was not conducive to improvement (H. Andersen, personal communication, March 16, 2014). His scholarship in the 1970s reflects his growing interest and teacher training in the implementation of inquiry, work to advance science education in developing countries, and exploration of relationship building between university and schools as a way to improve preservice science teacher education. Noticing that there was a need for discipline-specific sourcebooks to support high school science instruction, Hans invited some of his graduate students and junior faculty to co-author two books: *Sourcebook for Biological Sciences* (Troyer, Kellogg, & Andersen, 1972), and *Sourcebook for Earth Sciences and Astronomy* (Utgard, Ladd, & Andersen, 1972). These books are still in use today, illustrating the necessity and forward thinking involved in their development. In the 1980s, Hans became concerned with the lack of emphasis on science as a part of a basic education and the possibility of a general population incapable of participating in the scientific decision making that is necessary for a healthy democracy. In addition to publications on topics of science teacher education (i.e., nature of science, constructivism, and elementary teacher preparation) and co-authoring work with graduate students, Hans began to accept invitations to work on and lead committees at the state and national levels in an effort to be a change agent. In the 1990s and early 2000s, Hans participated in science education reform efforts. He co-authored the 1996 publication *Pathways to the Science Standards*, used by high schools to develop science curricula. His early enthusiasm for the National Science Education Standards and the attention science education received as a result of the reform efforts gave way to critique of the way they were being interpreted in some circles. He was frustrated by increased standardized testing and the amount of time being spent on preparation specifically for those tests, textbooks that "are a series of one-liners, with no elaboration, no central theme, and... boring," and "a 'layer cake' curriculum [which] leaves no room for students to explore or understand the underlying concepts or ideas. They simply memorize a jumble of seemingly unrelated facts and move on" (H. Andersen, personal communication, March 16, 2014).

Hans was a prolific writer. Over the course of his career, his passion and his work ethic led to authorship of eight textbooks and over 150 publications on science teacher preparation, science teaching, and science curriculum in state, national, and international publications. He wasa sustaining force for the Indiana journal, *The Hoosier Science Teacher*. Hans encouraged all students and faculty, past and present, to write for the publication. When submission rates were down, his close friend Nels Christiansen always came to the rescue. Strangely enough, Nels' writing style was similar to Hans, but satirical (H. Andersen, personal communication, March 16, 2014; J. Staver, personal communication, April 19, 2014).

Hans received numerous awards for teaching and mentoring over the years. His most coveted award was received from the class of 1970, when they identified him as one of six professors who positively influenced their lives. This poll was taken in 1980 by the IUB Alumni Office, a full 10 years after graduation (H. Andersen, personal communication, March 16, 2014).

There are several heartfelt and sometimes amusing stories that students and colleagues have shared about their experiences working with Hans. Carl Stedman (retired professor at Austin Peay State University, former dean of education and past-president of the Tennessee Association of Colleges for Teacher Education, commented, "[He] didn't simply provide knowledge, but [he] nourished the recognition of how to acquire it, which was much more important" (C. Stedman, personal communication, June, 2002). Eric Worch, associate professor of education at Bowling Green State University, said,

> The thing I remember most about [Hans'] secondary science methods course is [his] use of humor. [He] loved to show comics on the overhead projector. While I always got a laugh out of the comics, I got a bigger kick out of watching [Hans'] face after [he] unveiled a new transparency.... I suppose [he] had seen those comics many, many times over the years, but [he]

> laughed at them as if it were the very first time—a man truly in love with what he was doing. (E. Worch, personal communication, undated)

Carol Ann Franklin, professor of education at the University of the Redlands and director of educational technology, shared that Hans had been the "most influential educator (not counting my mom and dad, of course) in [her] life." Hans helped her get a position with Jim Winkleman at his old high school, Niles North High School. After a few years of successful teaching, she met up with Hans at an NSTA conference and he offered her a teaching assistantship and an opportunity to go on for her doctorate. She had spent many hours trying to make a decision about her future, and when she made her appointment with him he said, "I knew you would come back." It still makes her smile (C. Franklin, personal communication, September 2002).

Several individuals recalled some of the "Andersen Sayings." Larry Scharmann, the Melvin C. and Jane N. Nore Professor and Chair at the University of Nebraska-Lincoln, remembered Hans stating, "As much can be caught—as taught," and said that he caught it. Scharmann further explained that while Hans delighted in

> events, jokes, tricks and games that might have been left behind in junior high school. It allowed [him], nonetheless, to consummately serve undergraduates by letting them note that it was okay to be yourself, display a sense of humor, and appeal to the child that still hides within each of us. (L. Scharmann, personal communication, August 19, 2002)

Larry "cherishes" his students as much as he knows Hans did over the years. He added that Hans followed

> the advice of the Greeks—that there is beauty in simplicity. What need is there to memorize textbook facts concerning ecology and the environment, when the LORAX can make a more profound case? Likewise, [Hans] summed up Ralph Tyler's curriculum theory with crafty subtlety—"Interest is the point of departure in teaching and learning." (L. Scharmann, personal communication, August 19, 2002)

Larry "caught that one too." His undergraduate course philosophy starts with the "holistic model" that Hans shared with him and prepares them to respond to a question Larry's "New York friends would ask as [he] was growing up in New Jersey—'What's in it for me?'—as it applies to learning science."

James Dean Lubbers, environmental education specialist with the Missouri Department of Natural Resources and past president of the Missouri Environmental Education Association, wrote that while he was working on his MSES through SPEA, he met with Hans to talk about how to finish the six credit hours in science education to receive his life license in teaching. While there, Hans convinced him to enroll in the science and environmental education doctoral program, telling him, “You can roll your own marbles.” James continued,

> That line stayed with me for the two years my dissertation was incubating and was a significant motivator in finishing.... I have about ten years’ worth of marbles left to roll, even though I’ve lost a few along the way! (J. Lubbers, personal communication, September 2002)

Hans’ stories about his work in far-off locations stirred the imagination of many students. John Hansen, consultant, Hansen Research Services, LLC, told a story about leaving the safety of the nest at Indiana University to “lead a science department that didn’t exist” in Venezuela. He writes, “

> I stretched my wings, made sturdy from in-class practice labs where my excitement for exploring liquid nitrogen was met with [Hans’] words, “Here at the university we have many resources, it is just a matter of digging them up.” With this I dug through cultural boundaries and language difficulties in the small town of El Tigre to find dry ice in the ice cream man’s cart, German imported chemicals at the fine Italian leather good shop, and oxygen in the welder’s tanks in the back of the school’s warehouse (gymnasium). Having learned from my fateful fall during the egg drop practice-lab, I was able to swoop out of my lack-of-materials tail spin with students making anemometers from shoe boxes, cracking open little rocks with big rocks, and collecting flowers and insects in a mind boggling array of colors, shapes, sizes, fragrance, and degree of fearfulness.” (J. Hansen, personal communication, June 20, 2002)

Because of Hans’ international work, the Indiana University Science Education Department had a wonderfully diverse graduate student population. Students from Saudi Arabia, Pakistan, Thailand, and many other parts of the world amplified our shared experience in a positive way. Every now and then however, our colleagues needed a front lineman to run offense for them. Huey-Lien Kao, professor, Department of Natural Science Education, National Pingtung University of Education, wrote that he was very lucky to have Hans’ support and guidance in overcoming the difficulties associated with schoolwork and living in an unfamiliar place. He was grateful for the inspiration and knowledge that he and other international students received in the program. Huey-Lien recalled that a member of his committee didn’t think he could continue in the doctoral program because of his scores on the verbal portion of the GRE. Hans convinced this individual that while Huey-Lien was not fluent in spoken English, he was a bright student who would be able to write his dissertation. He continued,

> Every morning, early about seven o'clock, [Hans] worked at [his] office, looking through each student's work we submitted and returning us informative feedback promptly. Even during the continual process of revising doctoral theses, [he] never delayed any student's completion and presentation of theses. (H. Kao, personal communication, 2002)

In a similar line of thought, Don Troyer, retired professor of science education, Western Illinois University and namesake for the Donald L. Troyer Outstanding Mentor Award, wrote,

> What penetrated my thinking and life style was the way [Hans] consistently treated all who interacted with [him] with courtesy, respect and humility. [He] offered much to me in my professional studies and I thank you. But what [he] modeled for me as a human being was special and I attempted to emulate those behaviors in all my dealings with others during my years as a teacher and administrator. (D. Troyer, personal communication, August 18, 2002)

John Staver, professor of science education and chemistry at Purdue University, director of the Purdue University Center for Research and Engagement in Science and Mathematics Education, and past director of the Kansas State University Center for Science Education, was one of Hans' favorites. Both Hans and Sandy said he was like a son (H. Andersen, personal communication, March 16, 2014; S. Andersen, personal communication, August 21, 2014). John shared that like most undergraduate students, he struggled to find what he wanted to do in life. He was a premed who didn't have the grades required to get into medical school, and although he loved chemistry, he didn't want to work in a lab all day. Hans helped him understand that it wasn't a bad thing to not want to work in a lab and that there were plenty of other options for people who loved the subject matter, like teaching. John completed his undergraduate degree and taught at North Central High School for 6 years before coming back to IU to work on his PhD with Hans. John knows most of the stories of the jokes, tricks, and other special moments that occurred in the Science Education halls of distinction. He recalls stuffing helium balloons into faculty offices and the time when Hans pounded his first on his desk in undergraduate science methods class while stating, "You can't teach without behavioral objectives!" The story he likes to tell most is one about Hans' teaching methods. It seems that John was sick and missed his science methods class. When meeting Hans about how to make it up, Hans explained that he wanted students to begin to develop strategies for classroom management and fairness issues. Hans asked him to help set up a scenario where he would kick John out of class for being absent but let another student who was also absent stay in the class. Hans would be late and John would argue that the students should stay and wait for him and the other student would argue that everyone should leave because the professor wasn't there. John said, when Hans walked in about 10 minutes late, he put his stuff down on the front desk and said,

> Mr. Staver, it is so nice to see you. Did you have a good time playing pool? Did you lose track of time? You don't have to be here today if you don't want too. In fact, go. Leave. I'm sure you have better things to do.

John got up and left. The class was mortified. Even though Hans went out to get him and resume class, many of the students still thought that it wasn't an act. Everyone was engaged with the debriefing though (J. Staver, personal communication, April 19, 2014).

Hans was also known as a mentor and critical friend to colleagues both in his department and outside. Both Charlie Barman, retired professor of education at Indiana University-Purdue University-Indianapolis, and George Glasson, professor of education and program leader at Virginia Tech, credit Hans with influencing their careers. George Glasson recalled that Hans was teaching constructivism before most science educators knew what it was. During his brief tenure at IU, he said that what he learned from Hans was

> both useful and practical. [He] really taught me how to teach a science methods course in an inquiry manner (albeit pointing out the fallacies of "inductive reasoning"). From [Hans] I learned about discrepant events, hands-on/minds-on activities, the importance of prior knowledge, and connections of science and society. (G. Glasson, personal communication, August 28, 2002)

Charlie Barman wrote,

> [Hans has] always been helpful when I had questions related to advisement and science education in general. When [he] was president of NSTA, [he] made sure that I was given and assignment on the Awards and Recognition Committee and included me in other national activities. These activities helped me make some important contacts in NSTA that have been very beneficial. (C. Barman, personal communication, May 28, 2002)

Both Charlie and George mentioned the impact Hans had on them and many others through the many community-building efforts Hans organized, including the gatherings at "Andersen Acres." Bill Boone, professor of evaluation, measurement, and statistical analysis at Miami University, a close friend and colleague for 11 years, also cited the gatherings as a critical component in holding the program together and strengthening science education in Indiana (B. Boone, personal communication, September 18, 2002).

Service was a very important component of the triad for Hans. It was yet another opportunity to make a difference. He selflessly contributed his expertise to organizations at both the state and national levels. Hans became a fellow of the American Association for the Advancement of Science in 1967 and served as a member of the Electorate Nominating Committee from 1976 to 1979. He served on the Board of Directors for the Association for the Education of Teachers of Science (AETS) from 1977 to 1980 and as president from 1980 to 1981. In 1998, AETS presented Hans the Outstanding Mentor Award, their highest honor.

Hans was also active in the National Association of Science Teachers (NSTA). He served on the NSTA Board of Directors from 1980 to 1984 and later from 1998 to 1999 and as president from 1989 to 1990. During "The Hans Reign," there was increased attention on education. Some significant areas of policy that year were the Multicultural Division was established; the Board set a limit on how long a person could serve on the Board; NSTA supported the Peoples of the American Way in efforts to eliminate creationism in the science classroom; the National Science Foundation added elementary teachers to the Presidential Awards program; and the exciting but controversial alternative science program, The Scope, Sequence, 7 Coordination (SS&C) was up and running (Anonymous, personal communication, 2002). In 1992, Dr. Hans Andersen was given the Robert H. Carlton Award for distinguished service to science education. This is the highest award given by the NSTA.

In keeping his commitment to support improvement of science education at the state level, Hans served on the Board of Directors of the Hoosier Association of Science Teachers, Inc. (HASTI) from 1975 to 1978 and as president during the 2002–2003 academic year. In 1992, he was the recipient of the HASTI Distinguished Service to Science Education Award.

The International Connection

Hans Andersen was internationally renowned for his work in building science education programs here and abroad. In 1970, he was offered a position in Seoul, Korea but it was a war zone, and he and Sandy were afraid that it might escalate, so they decided against it. Later that year, four students from Thailand came to IU funded by the United Nations. The students went back full of ideas and new strategies, and Hans received an invitation to go to Thailand to help improve science education. Hans took a leave of absence from IU to become the UNESCO Senior Expert in Educational Techniques at the Institute for the Promotion of Teaching of Science and Technology (IPST) in Bangkok, Thailand. The Andersen family moved to Thailand in 1972 and formed lasting bonds with the people of Thailand (S. Andersen, personal communication, August 21, 2014). They called Hans *Ajahn,* a word that translates into "honored teacher." IU didn't offer leave of absences for longer than a year, so the family returned to Bloomington. However, that trip was followed by several other invitations to consult on projects to build science education programs in many parts of the world. Hans served as an Agency for International Development consultant to the Institute of Education and Research in Lahore, Pakistan; a visiting lecturer at Macquarie University in Sydney, Australia; a consultant to the University of the Witwatersrand in South Africa; and an appointee to the Commission of Higher Education in Papua New Guinea. Over the years, 12 Thai scholars came to IU to complete a PhD with Ajahn Hans and return to fill positions in academia. One of his students, Pussadee Tamthai, served as a Member of Parliament and then became the Governor of Bangkok. Hans also helped Thai students in other fields make connections to study in the United States. Following retirement, Hans returned to advise and give workshops at several Thai colleges and universities. In 2006, the Ministry of Education of Thailand awarded Hans an honorary medal of service for contributing to science education for over 40 years. In addition, he was granted an audience with the Crown Prince to receive an honorary doctoral degree from Pranakorn Rajaphat University in Bangkok.

Environmental Education

Hans Andersen's love for the natural environment played an important part in both his personal and professional life. He worked with Donald Winslow, IUB professor of education, to bring environmental education to preservice science teacher education and graduate students. Winslow was an early member of what is now called the Environmental Education Association of Indiana (EEAI) and encouraged Hans, which he did at the time that the name change took place. Hans was very proud of his EEAI membership card with the number 1. When I visited Hans to interview him for this chapter, he pulled his membership card out to show me that he still had it. Hans used the environment as a context for learning science but it wasn't until Project WILD came to Indiana in the 1980s that he was able to use it to the extent he wanted. Warren Gartner, conservation education supervisor of the Indiana Division of Fish and Wildlife, said that he was looking for an opportunity to get Project WILD curriculum to preservice teachers in Indiana. He contacted Hans and within the day, IU offered an independent study for between 1 to 4 credits for Project WILD training and practice. This served as a model later for Purdue and Indiana State University. In addition to the independent study, Hans began to use the lesson plans in his undergraduate methods courses and encourage graduate students who had taken the independent study to take the facilitators training. Warren said that the first time he met

Hans was at a facilitator's meeting at Bradford Woods in 1986. It was memorable because Hans was quiet and seemed to fit the stereotype of an academic. They started playing WILD Words (charades) and when it was Hans' turn, he drew a card and took off across the room swinging like a monkey. It surprised everyone. Warren knew that he was going to enjoy working with Hans. They held some back-to-back workshops (12 hours) for Project WILD and Project Learning Tree at IU, and Hans encouraged Warren to participate in the NSTA conference when it was in Indianapolis. They did "Oh Deer" in the Convention Center hall. Hans also took Project WILD to Thailand.

Hans was invited by IU President Myles Brand to serve on the Founding Board of the IUB Council for Environmental Stewardship in 1998, a precursor to the IU Sustainability Office. The Board was composed of deans, department chairs, students, and staff. It was an honor to be on the Council, certainly an innovative step in the academic world at that time. Although the Council met its demise 6 years later, it was able to establish the IU Teaching and Research Preserve at Griffy Lake. Hans was influential in helping to make this happen.

Mentors

Hans had many mentors along the way, but when writing about those who were most helpful in his professional career and "offered assistance and thoughtful guidance," he chose four (Hoosier Science Teacher, 2002, pp. 99–100).

Jim Winkleman, a teaching colleague in Skokie, Illinois, helped Hans distill his graduate science content rhetoric into smaller understandable bites for high school students. Jim Weigand taught him to appreciate and find joy in observing student learning. He also recruited him to do graduate work at IU and then asked him to stay. Willis Porter, an IU faculty member who was stationed in Bangkok, helped Hans when he visited Thailand for the first time. Willis introduced Hans to the *Sea of Smiles* found at the Bangkok Sunday Market and loathed certain administrators "with whom he became impatient." The last person he mentioned is Sandra, his wife, "Editor for Life, severest critic and best friend."

LEGACY

I had the privilege of talking with Hans prior to his passing. While his mind wasn't what it once was, he still had a pile of books to read on his desk, loved to talk about science and gardening, and still had an open-door policy for his "student."

Those who knew him best would agree that Hans' legacy resides in the people, students, and colleagues whom he came in contact with during his life. IU colleague and friend, Frank Lester, retired professor of mathematics, wrote,

> He inspired generations of budding science teachers to be the best teachers they could be. [He] also inspired generations of science education doctoral students to become the best they could be. Without exception, these students regarded [Hans] as a model of what a university science educator should be. (F. Lester, personal communication, September 24, 2014)

Diana Hunn, associate professor of education at the University of Dayton and executive director emeritus of the Science Education Council of Ohio, stated that Hans had been a "model teacher and learner."

> I find myself leading discussions that mirror many I witnessed in [his] classes and hoping that the undergraduates gain knowledge and skills as [his] have in the past.... [Hans'] national presence with NSTA [was] outstanding, and [his] dedication to the grassroots folks at the local and state levels made a difference in the lives of teachers every day. What a wonderful legacy of learning. (D. Hunn, personal communication, September 2002)

George Ladd, professor of education, Boston College, stated that he could "honestly say that each day of my post-Indiana career I reflect on (Hans'] model as I interact with my own students" (G. Ladd, personal communication, June 21, 2002). Peter Rubba, professor emeritus of education and retired director of academic affairs for graduate programs, Penn State World Campus, credited Hans' "mentoring in research and teaching" for his "success as a science educator" noting the Nature of Scientific Knowledge Scale (NSKS) is still used worldwide. "The measure of a career is not any one student, but the many students at all educational levels that you have touched, the students of those students, and the generations of students who follow" (P. Rubba, personal communication, September 1, 2002).

When I asked Gerald Krockover, professor of earth and atmospheric science education at Purdue University, what he thought Hans' legacy was, he said that his legacy is the "number and quality of science teachers and science educators that have had the opportunity to learn from him. Hans was one of the first people to welcome me to the science education community and that community existed because of him" (G. Krockover, personal communication, September, 2014). Hans was a connector and community builder, and the relationships that resulted were the power in strengthening science education. When the science educators from all of the universities worked together, they were able to find a common voice to work with the state and make policy. "You just don't see that anymore." Gerry also said that "Hans served as the inspiration for his students to take leadership. Carolyn Hayes, a former student of Hans', is now the president-elect of NSTA"

(G. Krockover, personal communication, September, 2014). Carolyn, also the statewide curriculum specialist, Indiana School of Medicine, agreed.

> He did serve as my role model. In fact, I remember the theme from his NSTA presidency, "Pride Inside." He wanted teachers to be proud of the work they did. That they are professional. That the work they do is honorable and that they should be proud of their accomplishments. (C. Hayes, personal communication, October, 2014)

She was so impressed with the theme that she has used it in a couple of different venues, one of which was her keynote address at the Indiana President's Awards Luncheon. Carolyn says that she "wants them to have 'Pride Inside'—to tell their story. I'm going to extend Hans' legacy" (C. Hayes, personal communication, October 2014).

As I closed the interview with Hans' and gathered my things, I asked him, "What do you think your legacy is?" He sat back in his chair and with a twinkle in his bright blue eyes, he said, "I know what it is. It's what's inside my students. It's you" (H. Andersen, personal communication, March 16, 2014).

Hans' passed away on June 4, 2014. But, he lives on inside each of us who were fortunate to be able to call him our teacher.

ACKNOWLEDGMENTS

Thank you to all of Hans' many friends and colleagues who generously sent letters of congratulations to be included in the book of tribute for Hans O. Andersen on the occasion of his retirement in September, 2002. I am also grateful for Hans' family and friends who shared special memories of Ajahn Hans for the purposes of this chapter.

REFERENCES

Andersen, H. O. (1969). Science methods instruction. *American Biology Teacher, 31*(4), 250–251.

Andersen, H. O. (2002). Hoosier science teacher. *My Mentors, 32*(5), 99–100.

Andersen, H. O. & Winkleman, J. (1962). An approach to team teaching in biology. *American Biology Teacher.* Retrieved from http://www.jstor.org.proxyiub.uits.iu.edu/stable/4440129?seq=1#page_scan_tab_contents

Andersen, S. O. (2014, July). [Web log post]. Retrieved from https://scottoandersen.wordpress.com/2014/07/

Indiana University. (2002). *Faculty vita. Accreditation report 2002.* Retrieved from http://www.indiana.edu/~iuncate/facultyvita/andersen.html

Institute for the Promotion of Science and Technology. (n.d.) *In memory of Hans O. Andersen.* Retrieved from http://www.ipst.ac.th/eng/index.php/news/news-and-announcement/item/24-in-memory-of-professor-hans-o-andersen

Troyer, D. L., Kellogg, M. G. & Andersen, H. O. (1972). *Sourcebook for biological sciences.* New York, NY: MacMillan.

Utgard, L., & Andersen, H. O. (1972). *Sourcebook for earth sciences and astronomy.* New York, NY: MacMillan.

Weigand, J. (2002). *Hans O. Andersen. Tribute given at the 2002 Indiana University retirement dinner.* Indiana University, Bloomington, IN.

Dissertations Hans Andersen Guided With Handwritten Inscriptions Included

1967–1969

McFee, Evan E.: *The Relative Merits of Two Methodologies of Teaching the Metric System to Seventh Grade Science Students.* (September 1967)

Baptiste, Hansom Prentice, Jr: *The Effect of an Equilibrated Methodology on the Acquisition of the Concept-Conservation of Quantity.* (September 1968)

> "To Hans: A true friend to teachers. From Prentice"

Ferrence, Gary M.: *The Quantification and Qualification of Verbal Communication in the High School Biology Laboratory.* (September 1968)

Koutnik, Paul George: *The Effect of Cognitive Style, Sex, and Degree of Completeness of an Electric Data Background on Student Prediction Of Circuital Function.* (September 1968)

Mansla, John Anthony: *The Effects of Instruction In Interaction Analysis on the Verbal Inquiry Patterns of Elementary Science Methods Students.* (September 1968)

Williams, Kenneth Edwin: *The Feasibility of the Use Of Dimensional and Unit Analysis in Junior High School Science Instruction.* (September 1968)

> "With grateful appreciation. Ken"

Taylor, Paul A.: *Concept Learning Using Positive and Negative Instances in Learning the Classification Scheme of Bloom's Taxonomy.* (December 1968)

Foss, Harold F.: *Identification of Laboratory Features in the First Course of Microbiology in Various Kinds of Colleges and Universities.* (June 1969)

Konetski, Louis C.: *Instructions Effect on Questions Asked by Pre-Service Science Teachers.* (June 1969)

> "Hans,
>
> It was both an honor and a pleasure (though sometimes frustrating because of my lack of understanding) to work with you for two years. No one has contributed more to my knowledge and potential creative thought in the field of science education and in sensitivity to others. If I am successful, I owe it to you; if I am not, I will have only myself to

> blame. No one could have been more helpful in successfully guiding the first Polack through the science ed. Program at I.U.
>
> Thank you for everything—I will always respect you for what you want to do and for the potential that you have. I hope that you will continue to work with me for bigger and better things.
>
> I'll keep in touch.
>
> Louis"

Trojcak, Doris Ann: *Five Stages of Instruction for Sequencing Science Activities According to Gagne's Learning Model.* (June 1969)

> "Dear Hans, dear Hans dear—
>
> My pre-dissertation criterion for judging the goodness of a person was if he could smile at Blepo. My post-dissertation criterion is if he can accept, persevere, and end up smiling as Chairman of a dissertation. You, Hans, are a doubly good man—and I'm forever grateful. Doris Trojcak"

Wise, Lew Edgar: *The Relative Merits of Two Methodologies Designed to Improve the Eighth Grade Student's Ability to Recognize the Relevancy of Data to Selected Problems in Science.* (June 1969)

Hittle, David R.: *Directed Discovery, Form of Presentation, and Laboratory Schedule for Non-Science Students in the Freshman Chemistry Laboratory.* (September 1969)

Ladd, George T.: *Determining the Level of Inquiry in Teachers' Questions.* (September 1969)

Troyer, Donald Lee: *Conceptual and Factual Educational Set of Teacher and Learner on Performance in the BSCS Curriculum.* (September 1969)

Utgard, Russell O.: *Verbal Behavior of Recitation Teachers and Achievement of College Geology Students.* (September 1969)

1970–1979

Falls, William R.: *The Effects of Research Papers and Reports on Understanding Science When Used in Secondary School Physics.* (June 1970)

Shaver, John D.: *The Effect of Laboratory Teaching Strategy on Pupil Task Orientation.* (June 1970)

> "For all your time and energy spent in my behalf, my sincere thanks. John D. Shaver"

Ault, Frederick Keith: *Cognitive Style, Attitude toward Science, and Sex on Success with Programmed Instruction of Kinetic Theory.* (September 1970)

Bryant, Napoleon, Jr: *The Effects of Performance Objectives on the Achievement Level of Selected Eighth-Grade Science Pupils in Four Predominantly Black Inner City Schools.* (September 1970)

"Hans: For the last 2 years you have been my advisor, mentor, and friend. Of the three, the latter by far will endure through the many years that lie ahead. Napoleon"

Garren, Donald R.: *Effects on Achievement when Excerpts from Physics are Interjected into Programmed High School Biology.* (September 1970)

Godomsky, Stephen F.: *Programmed Instruction, Computer-Assisted Performance Problems, Open-Ended Experiments and Student Attitude and Problem-Solving Ability in Physical Chemistry Laboratory.* (September 1970)

Mills, Terence John: *Prospective Elementary Teachers' Problem Solving Behavior, Effects of Critical Thinking, Cognitive Style, and Preferred Teaching Level.* (September 1970)

Rusch, John Jay: *Investigation of an Instructional Treatment Designed to Alter Reacting Behaviors of Prospective Secondary Science Teachers.* (September 1970)

Stekel, Frank D.: *The Effects of Increasing the Student Involvement in a College Physical Science Laboratory Program.* (September 1970)

Hjelmhaug, Noel Nels: *Context Instruction and the Ability of College Students to Transfer Learning.* (June 1971)

Funk, H. James: *Test-Construction Behavior of Pre-Service Secondary Science Teachers.* (September 1971)

Rezba, Richard J.: *Preparation of Pre-Service Science Teachers in the Use of Alternate Laboratory Teacher Behaviors.* (September 1971)

"Hans, A pontifical cascade of superlatives would not suffice to express my gratitude for your help and understanding, or as we southern folk say, 'Thank you.' Dick"

Black, Thomas Rile: *An Investigation of Levels of Questioning Found on West African School Certificate Science Examinations and Nigerian Science Teachers' Examinations.* (November 1971)

Adhikary, Poorna Kanta: *Instructional Effect on Question Asking Behavior of Prospective Peace Corps Science and Mathematics Volunteer Teachers.* (May 1972)

Strahler, Violet Ruth: *Relationship of Eight Predictive Factors to Success in Ninth Grade Science and Mathematics.* (May 1972)

"To Hans Andersen with high esteem and sincere appreciation. From Violet (April 1972)

Souers, Charles V.: *Incidental Treatment Effects on Student Attitudes toward a Current Social Problem.* (May 1972)

Doyle, Patrick J.: *An investigation into Factors Effecting the Learning and Utilization of a Questioning Technique by Elementary Methods Students.* (August 1972)

Fiel, Ronald L.: *An Investigation OF The Effectiveness of Formative Evaluation and Remediation in Achieving Mastery of Intellectual Skills.* (August 1972)

Smith, Walter Scott: *The Effect of Alternative Types of Instruction on the Learning of a Question-Asking Skill by Pre-Service Secondary Science Teachers.* (July 1973)

Burrows, Charles K.: *The Effects of a Mastery Learning Strategy on the Geometry Achievement of Fourth and Fifth Grade Children.* (August 1973)

Ramey, Carl Victor: *The Identification and Remediation of Mathematical Skill Deficiencies in Freshman College Chemistry.* (August 1973)

Bogut, Thomas L.: *The Effects of Sequencing Selected Instructional Strategies on the Attitudes and Open-Mindedness of Pre-Service Elementary Science Teachers.* (May 1974)

Bedwell, Lance Eugene: *The Effects of Training Teachers in Question-Asking Skills on the Achievement and Attitudes of Elementary Pupils.* (August 1974)

Repicky, Paul Albert: *The Effects of Feedback from Systematic Observation on the Performance and Attitudes of Pre-Service Secondary Science Teachers.* (August 1974)

Knight, Linda Bernhard: *The Effect of an Early, Field-Based Experience in ISCS Classrooms on the Attitudes and Behaviors of Pre-Service Science Teachers.* (August 1975)

> "Hans—Your leadership and friendship has been a cornerstone in my life. I can find no words to adequately express my gratitude. Thank you. Linda."

Thomas, Frederick James: *An Exploratory Study on the Cultural Relevance of Science Instructional Materials: The People's Republic of China.* (August 1976)

Suwanarat, Pimshai: *The Effects of Pictorial Stimuli on Learning Selected English Words by Thai Fifth Grade Children.* (January 1978)

Jantaraweragul, Pote: *Relationships among Cognitive Style, Attitude toward Science Activities, and Science Process Skills.* (March 1978)

Suchareekul, Jariya: *The Inquiry Behavior Preference and Performance of Thai Science Teachers.* (May 1978)

Clark, Margaret Mary: *An Examination of Metric Performance and Learner Aptitudes of Elementary Pre-Service Teachers Using Computer-Assisted Instruction.* (Augusts 1978)

Jariyavidyanont, Sagol: *Job Satisfaction of Nida Faculty Members.* (August 1978)

Lucas, Keith Bramwell: *Cognitive Structure Changes in Astronomy Students Using Advance Organizers Matched with Cognitive Style.* (August 1978)

Staver, John R.: *The Development and Construct Validation of a Group Administered Test of Piaget's Formal Thought.* (August 1978)

> "Hans, Only a very few people exert a major influence on a person during his life. You have been, are, and will continue to be such a person for me. John"

Hassan, Hassan A.: *Pupil Control Ideology and the Science Classroom Environment in the Secondary Schools of Khartoum, Sudan.* (December 1978)

> "To: Hans Andersen. For his unlimited support and regard... Hassan"

Shimer, Stanley Stuart: *Pre-service Elementary School Teachers' and Science Education Experts' Perceived Values for Selected Science Methods Curricular Topics.* (May 1979)

> "To Hans Andersen. Thank you for the leadership, patience and friendship. 1979 Stan S3"

Eggins, Judith Anne: *The Interaction between Structure in Learning Materials and the Personality Type of Learners.* (July 1979)

Rowe, John David: *The Influence of Training in Textbook Selection Procedures on Pre-Service Teachers' Perceptions and Use of Text and Curriculum Materials.* (August 1979)

> "To Dr. Andersen, With heartfelt thanks for your many kindnesses, and contributions to what has been a rewarding and enjoyable sojourn at Indiana University and in the United States. John Rowe"

1980–1989

Positong, Pollasanha: *Facilitating the Development of Inquiry Teaching Skills through Rating and Practice.* (May 1980)

Sherwood, Robert D.: *The Effect of Selected Instructional Strategies on the Problem Solving Ability of High School Chemistry Students as Related to Their Proportional Reasoning Ability and Verbal-Visual Preference.* (May 1980)

> "Hans, I really cannot express the thanks I owe to you in words. Please let me know how I can be of service to you in the future. Bob Sherwood"

Omar, Enayat Shaker: *Effects of Training Pre-Service Science Teachers with Respect to Their Preference for and Performance of Inquiry Behaviors in Relationship to Open or Closed Mindedness in Egypt.* (January 1981)

Al-Faleh, Nasser A.: *Effects of Lecture-Demonstration and Small Group Experimentation Teaching Methods on Saudi Arabian Students' Chemistry Achievement and Attitudes toward Science Learning.* (April 1981)

> "To: Dr. Hans Andersen, Great appreciation to his help, support, encouragement and unlimited advises in getting this study completed. Nasser A. Al-Faleh"

Franklin, Carol Ann: *Instructor Versus Peer Feedback in Microteaching on the Acquisition of Confrontation; Illustrating, Analogies, and Use of Examples; and Question-Asking Teaching Skills for Pre-Service Teachers.* (June 1981)

Tashkandi, Mohammad Omar: *Effects of Instruction and Personal Traits of Saudi Pre-Service Teachers on the Use Of Higher Cognitive Questions.* (June 1981)

Tamthai, Pusadee Piyakul: *The Effect of Advance Organizer on Science Learning Achievement of Eighth Grade Thai Demonstration School Students with Average Academic Ability.* (December 1981)

> "To 'khru' Hans, A delightful person, a wonderful teacher and a very dear friend who introduced me to this fine university and the peaceful town of Bloomington. Love, Pusadee Tamthai—Sept. 28, 1982)"

Enochs, Larry Gene: *Implementation Proneness in Terms of Teacher Factors Relating to Inservice on Selected Science Education Trends: A Case Study.* (January 1982)

Brooks, Edwin, T.: *The Effects of Mastery Instruction on the Lea Ring and Retention of Science Process Skills.* (May 1982)

Ilyas, Mohammad: *Relationships between Science Process Skills Instruction and Secondary School Teachers' Performance, Use and Attitudes toward Using These Skills.* (January 1983)

Al-Mansour, Abdulaziz Mohammad: *Attitudes of Elementary Teachers in Saudi Arabia toward Teaching Science in Grades 4-6.* (August 1983)

El-Rashes, Ibraheem Mohammad: *Influence of Teacher Wait-Time and Question Level on Eighth Grade Students' Science Achievement and Attitudes toward Science in Saudi Arabia.* (January 1984)

> "To Pro. Hans Andersen whose ideas, efforts never failed me. I thank you for devoting part of your time for this study. I. Rashed"

Hamrick, Linda: *Influence of Re-Sequencing General Science Content on the Science Achievement, Attitudes toward Science and Interest in Science of Sixth Grade Students.* (May 1984)

> "To: Dr. Hans Andersen, May 1984. Hans—Thank you for your academic guidance, professional encouragement and personal support. It has been my honor to have had you chair my program committee and to have been your student. Perhaps someday my professional accomplishments may merit the investment of 3 years of steadfast effort on your part. I remain gratefully, and with… Best regards always, a product of your guidance—Linda"

Samuel, K. V.: *A Longitudinal View of the Influence of a Coordinated Science Methods Course Sequence on Pre-Service Elementary Teachers' Pupil Control Ideology and Locus of control.* (August 1984)

> "Dr. Andersen: Thank you for all your thoughtful assistance during my studies here and especially in the course of this project. Sam"

Al-Hemaisan, Mohammed Ibrahim: *Science Achievement, Attitudes toward Science, Learning Motivation, and Divergent Creativity of Saudi Arabian Middle School Male Students Identified as Academically Gifted or Non-Gifted.* (August 1985)

Scharmann, Lawrence Conrad: *The Effects of Sequenced Instructional Strategy and Locus of Control on Pre-Service Elementary Teachers' Understanding of the Nature Of Science.* (August 1985)

> "Hans, I wish there were some way to adequately express my most sincere appreciation for your service, guidance, and friendship these past three years. Don McCurdy did steer me correctly in suggesting I.U. Thank you for everything! Larry Scharmann

Chiang, Shing-Ho: *Individualized Instruction for Junior High School Physics in Southern Taiwan.* (August 1986)

> "Dear Professor Andersen: I'd like to dedicate this thesis to you for representing my deepest appreciation. Shing-ho Chiang—Aug. 14, 1986"

Hunn, Diana Marie: *Self-Concept in Science Teaching and Comparisons of Pre-Service Elementary Education Majors in Field-Based and Traditional Science Methods Programs.* (August 1986)

"Hans—With deepest thanks for your advice, counsel, and continued friendship over the years. Your example has made lasting impact on my life—personally and professionally. I hope you know how much your dedication has meant to each of your students. I certainly hope to repay some enthusiasm for science education. Diana"

Purepong, Wichit: *Relationships between Thai Pre-Service Teacher Sciencing Self-Concept and Five Teaching-Learning Affective Attributes.* (May 1987)

"Dr. Andersen: Thank you very much for everything. You're the very best teacher I've ever had. Wichet"

Leou, Shian: *A Comparison of the Effects of Individualized Versus Conventional Instruction on the Mathematics Performance of Seventh Grade Students in Southern Taiwan.* (July 1987)

1990–1999

Lessow, Barry D.: *Factors related to Elementary Teachers' Effective Utilization of Field Trips to Informal Science Resources.* (October 1990)

Lessow, Denise E.: *Visitor Perceptions of Natural Habitat Zoo Exhibits.* (December 1990)

Lai, Dhing-San: *An Evaluation of Elementary School Teachers' Science Preparation in Taiwan.* (May 1992)

Baird, Kate, A.: *Behind Observable Behaviors: Model Indiana Biology Teachers Speak Out on Students Evaluation and Assessment.* (May 1994)

Bowman, Jacqueline K.: *The Use of Mnemonics in Constructivist Teaching.* (May 1994)

Svec, Michael T.: *Effect of Microcomputer-Based Laboratory on Students' Graphing Interpretation Skills and Conceptual Understanding of Motion.* (June 1994)

Sirochman, Rudy R.: *A Quantitative/Qualitative Study of Conceptual Change in a College Physics Class.* (August 1996)

Blank, Lisa M.: *Metacognition and the Facilitation of Conceptual and Status Change in Students' Concepts of Ecology.* (June 1997)

"Hans, Thanks for an incredibly satisfying & challenging 3 years... I couldn't have done it w/out your guidance and support. Cheers, Lisa"

Page, Scott M.: *A Case Study of an Outdoor Environmental Learning Center at an Elementary School.* (July 1997)

"Hans, Thanks for all of your help. I wouldn't be where I'm at today without it. Scott Page"

Kao, Huey-Lien: *Case Study of Science Teaching in an Elementary School—Characteristics of an Exemplary Science Teacher.* (March 1998)

"Hans, Thanks for all of your help. I would like to dedicate this thesis to you for representing my deepest appreciation. Huey-lien"

Worch, Eric Alan: *Play in Four Species of East African Monkeys: Implication for Early Childhood and Elementary Education.* (August 1998)

> "Hans, Thanks for bearing with me on this one. Not everyone has ability to detect the importance of basic research in education. You're the greatest. Eric"

Liang, Ling L.: *The Effects of a New Constructivist Science Curriculum (PIPS) for Prospective Elementary Teachers.* (May 1999)

Saam, Julie Reinhardt: *A Case Study of a Mathematics Teacher's and Science Teacher's Use of Teacher Wisdom in Integrating Middle School Mathematics and Science Content.* (October 1999)

2000–2002

Khalid, Tahsin: *Pre-Service Elementary Teachers' Misconceptions with Respect to Three Environmental Issues.* (March 2000)

Nowak, Jeffrey A.: *The Implications and Outcomes of Using Problem-Based Learning to Teach Middle School Science.* (April 2000)

Cakiroglu, Jale: *Pre-Service Elementary Teachers' Self-Efficacy Beliefs and Their Conceptions of Photosynthesis and Inheritance.* (May 2000)

Petty, Michael Eugene: *A Case Study of the Abductive Reasoning Processes of Pre-Service Elementary Education Students in a Role Playing Setting Concerning a Mock Senate Hearing on Global Climate Change.* (September 2001)

MaKinster, James G.: *The Effect of Social Context on the Reflective Practice of Pre-Service Science Teachers: Leveraging a Web-Supported Community of Teachers.* (May 2002)

> "Hans, Thanks again for everything. Jim"

Yoon, Jiyoon: *Feasibility and Effects of Science Education for Parents through the Internet (SPI) in Korea.* (May 2002)

> "You are my forever teacher and mentor. Thank you Hans! 6/25/02 Jiyoon Yoon"

CHAPTER 9

VINCENT LUNETTA

A Pioneer and Advocate for Science and Technology Education and for International Collaboration

Patricia Simmons
North Carolina State University

The greatest blessing of my professional life: The students and colleagues with whom I've been so fortunate to work and grow.

—Vincent Lunetta (personal communication, 2014)

Dr. Vincent Lunetta began his career as an educator teaching high school physics and physical science for 10 years while also serving as science department head. During his early years of teaching students science, he distinguished his career pathway as one of the earliest educators (in the 1960s) to develop, use, and assess computer simulations in high school science learning. (He used that work, in part, as a tool to help his science department acquire computer time-sharing at little cost.) His budding interest in the use of technology in science teaching led to a career-long focus on appropriate uses of technology in the learning and teaching of science at all levels. A second career-long focus that emerged in Lunetta's first decade as a science

Going Back for Our Future II, pages 153–164

teacher focused on the important roles and nature of school laboratory investigations and projects from the 1960s through the present day.

Lunetta's academic career spanned three decades in higher education at the University of Iowa and Pennsylvania State University. His contributions to science education led to the authorship of 200-plus articles, books, and computer simulations. Several of his computer-based simulations in science education and in teacher education were widely distributed and offered innovative, interactive simulations for use in introductory science courses and in teacher education programs. Throughout his career, he worked with K–12 and university teachers to engage their students in learning science through the active exploration of science concepts and applications visible in their local and regional school environments and cultural contexts. He used contemporary technologies as important elements of inquiring, data gathering, and data analyses in innovative science lessons. Lunetta taught and/or mentored approximately 1,000 graduate students and served on many dissertation committees. His contributions to science teacher education, goals for science learning, the nature and roles of school laboratory investigations and the roles of technology applications in developing learners' interest, concepts, and problem-solving skills, have influenced science teachers and science education scholarship throughout his career and into retirement.

AT THE FOREFRONT OF TECHNOLOGY, SCIENCE INQUIRY, AND TEACHING

During summers in his college years, Lunetta is proud to say that he worked for the Mount Washington Cog Railway, the first cog railway in the world. Happily, his work did not stop with the steam engines there! After graduating from Harvard College and 2 years of active service as an Army officer, Lunetta's early teaching experiences included his using and adapting science curricula developed in the post-Sputnik era. At his first National Science Teachers Association (NSTA) meeting, he was moved by a presentation given by Linus Pauling entitled, "The Social Responsibility of Scientists." This pivotal experience led him to initiate an extracurricular activity with high school students to help them connect the sciences they were studying with history, philosophy, and the arts. This collaboration with his students led to his first *STAR* award from NSTA, one of his first publications in science education, and career-long attention to important intersections of science teaching with cultural contexts.

During his 10 years of service as a science teacher, Lunetta was involved in teacher professional development as well as in the design and implementation of a science wing in a new high school. Delving into experiences that would help teachers become more effective led him to the University of Connecticut, where he worked on a doctoral study associated with the design, development, and classroom trials of simulations and computer software supporting meaningful science learning. Lunetta described his now lifelong interest in technology (with special attention to simulations and interactive graphics and the nature of science) as a great opportunity in a time of sea change in technologies and in worldviews about learning and teaching:

> Early in the advent of time-shared and later personal computers in schools, I felt that computer technologies could play important roles in helping students organize and visualize data, raise questions, and explore the effects of variables and accuracy of measurement. I also felt that while appropriate lab investigations with science materials are essential in comprehensive science learning, appropriate experiences with simulation of science phenomena could also play important roles in enhancing the depth and breadth of students' understanding. As personal computer technology developed, as real-time digital data gathering became available in schools, and as interactive graphics emerged as a particularly important learning tool, my interest in the important role of appropriate computer technologies in learning and teaching science increased substantially. (Lunetta, personal communication, 2014)

His first full-time faculty appointment at the University of Iowa led to becoming Director of Iowa-UPSTEP, a model program for science teacher education that was recognized for exemplary practices. Working closely with colleagues at Iowa and with mentor teachers in schools, prospective teachers in Iowa-UPSTEP engaged in innovative, model-based experiences in their clinical experiences in schools and in their professional development in science teaching. At the time of his initial appointment at Iowa, he discovered that several science teachers in Iowa City, Cedar Rapids, and nearby school districts were teaching with "Self-Paced" science curriculum materials. As he and the teachers worked together, several classrooms became excellent sites for the science teacher interns to engage with "PACE" students, in what became a series of "clinical experiences" in school classrooms. For these experiences, Lunetta and his colleagues developed specific tasks the preservice teacher interns could use to better listen to and understand their students' conceptual understandings and alternative concepts. The clinical experiences were designed to help the teacher interns hone their own questioning, listening, and responding-teaching skills. Early clinical experiences in the Iowa-UPSTEP program for prospective secondary school science teachers also included working with students in elementary school classrooms to gain better understanding of the needs of learners at various stages of their science learning and development. Lunetta's understanding and sensitivity to teachers' time constraints, school expectations, and classroom realities enabled him, his colleagues, and collaborating teachers to develop and employ clinical experiences for prospective teachers that supported students' classroom science learning, while simultaneously developing professional skills for the preservice teachers.

The initial foci for Lunetta's scholarship were shaped by the principal task for which he was hired—the development, leadership, and teaching in the newly established Iowa-UPSTEP program for the preparation of secondary school science teachers. His leadership role was to provide special expertise in activities associated with teaching in the physical sciences. As these tasks and scholarship evolved, he sought important input from students, collaborating teachers, and colleagues to strengthen and shape the role for outstanding secondary science teacher education programs. Six initial foci that Lunetta, his colleagues, and graduate students employed in designing, implementing, assessing, and modifying the Iowa-UPSTEP program included

1. Reviewing goals for the knowledge and skills of secondary school science teachers with special attention to goals articulated in the contemporary NSF-sponsored secondary science curriculum projects of that era;
2. Becoming better informed about contemporary science teacher preparation programs with excellent reputations and related scholarship;
3. Communicating with local and regional science teachers to understand their goals for students' learning and their perceptions of accomplishments, challenges, opportunities, and needs;
4. Articulating Iowa-UPSTEP program goals for science teacher learning and developing an organizing model for UPSTEP program development consistent with those goals;
5. Developing and implementing experiences and activities to promote UPSTEP learning goals in science teacher education, respond to needs of the prospective teachers and their students, and respond to needs of collaborating teachers; and
6. Developing and assessing achievement of program goals with special attention to the learning and development of the prospective science teachers.

The effects of specific program experiences on the prospective teachers, on the students they were teaching and studying, and on their mentor teachers, were examined by faculty and graduate student staff. This scholarship resulted in adjustments in program practices and in professional development in science teacher education for associated graduate students and faculty.

At the University of Iowa, Lunetta pursued interests about the learning and teaching of physics and the physical sciences, and on activities and practices teachers could implement with their students to promote desired learning outcomes. His attention to supporting science teacher education can be observed in his book, *Inquiring and Problem Solving in the Physical Sciences: A Sourcebook* (Lunetta & Novick, 1982).

> Most science teachers want to help their students improve inquiring and problem-solving skills in the course of studying science... teachers will find numerous practical suggestions... [to] use the activities with large groups, small groups, or individual students... will find some answers to questions regarding the relevance of inquiring and problems solving to goals for science teaching, to the nature of science, and to the ways people learn. (p. vii)

An early seminal publication in teacher education was *Classroom Transactions,* a first ASTE (Association for Science Teacher Education, formerly AETS) print publication of the NSTA (Lunetta & Zalewski, 1974). The publication presented numerous interesting but problematic science lab and classroom incidents that often confront early career science teachers. These incidents and scenarios were used in teacher preparation classes to stimulate discussion of the effects of varied teacher responses, to promote consideration of problem resolving strategies, and to discuss similar problems the prospective teachers had observed. This publication later became the basis for a series of widely distributed interactive computer simulations in which the classroom-problem situations evolved based upon the sequence of teacher behaviors selected by the prospective teacher. Problem development, further challenges, and crises and/or problem resolution were based upon a human transaction model that was popular and contemporary when these teaching simulations were in use.

As a science teacher, Lunetta became intrigued with the powerful effects of appropriate laboratory experiences on student interest and understanding. He noted, in particular, how his students could become very interested and engaged in some school laboratory investigations and projects. He also observed that the pursuit of appropriate investigations appeared to help students understand scientific procedures as well as science concepts. At the University of Iowa, Lunetta pursued this interest with his graduate students and faculty colleagues by examining roles of teachers, curricula, and classroom-laboratory resources in engaging students in meaningful investigations in the school laboratory and in students' projects. This led to an early series of publications examining the investigative behaviors students would perform if they and their teachers followed the instructions printed in widely used textbooks and teachers' guides. Four of the papers in this series included "An Analysis of Laboratory Inquiries in the BSCS Yellow Version" (Tamir & Lunetta, 1978); "An Analysis of Laboratory Activities: Project Physics and PSSC" (Lunetta & Tamir, 1981); "Do Secondary School Laboratory Texts Reflect the Goals of 'New' Science Curricula?" (Fuhrman,

Lunetta, & Novick, 1982); and "Matching Lab Activities with Teaching Goals" (Lunetta & Tamir, 1979).

Later in his professional life and after retirement, he authored and co-authored several widely cited review chapters on the role of the school laboratory in science education. A quotation from the widely cited article, "The Laboratory in Science Education: Foundations for the 21st Century" (Hofstein & Lunetta, 2004), highlights Lunetta's career-long concern regarding the central role science teachers play in helping diverse students learn and appreciate science. It also reflects his efforts to promote scholarship that will inform and shape teaching goals and practices.

> It is important to provide support for teachers (including time and opportunities) to collaborate with colleagues in the science education research community so as to understand, develop, and teach in ways that are consistent with contemporary professional standards and research on learning. Competent professional teachers have important roles to play in the continual renewal and development of science education standards and in supporting and doing related classroom-based research that can shape science-teaching practices. Empowering professional teachers in these roles and encouraging relevant research on central issues like supporting and assessing the effectiveness of the school science laboratory are very important next steps that warrant attention from professional societies, higher education, school administrators, and teacher certification bodies. (p. 58)

INTERNATIONAL COLLABORATIONS IN SCIENCE EDUCATION

Thanks to the leadership of Dr. Robert Yager, the visits of international science educators, and substantial numbers of international students at the University of Iowa Science Education Center, Lunetta had numerous opportunities to explore and develop interests in international science education. One of the most important contributions of Lunetta's career has been his role working with students and scholars in diverse international settings. Among other things, he was able to promote applications in the sciences relevant to students in their unique local and regional environments. In Australia, for example, a joint article showed the importance of including marsupials in Australian biology school textbooks (DeLaeter & Lunetta, 1982), and in Latin America on relevant local biology curriculum (Lunetta & de la Rosa, 1983).

On many occasions at the University of Iowa and at Penn State, he dedicated much of his time to mentor and assist international graduate students. Many students remember well the lines of international students convening outside Lunetta's office to talk with him about their careers as well as their perceived disconnects associated with being in a "strange" new country with schooling very different from their own experiences. These talks often spanned well into evening hours as he selflessly mentored and listened to students. In the process, professional and personal relationships developed with former students and colleagues in several countries outside the United States leading to long-term collaborations across countries, including Australia, Brazil, China, Germany, Greece, Indonesia, Israel, the Netherlands, New Zealand, the Philippines, Portugal, South Africa, Taiwan, Turkey, and the United Kingdom. These collaborations resulted in coauthored monographs and scholarly publications for science educators in universities and local schools. His earliest co-authored publications generally addressed contemporary issues intersecting school science, the role of investigations in science learning, goals for science learning, and the roles of technologies in learning and teaching science. Some later publications addressed matters of science education policy, such as "Reform of Science Education in Taiwan: Contexts, Opportunities, and Concerns" (Lunetta & Lederman, 1998).

INTERSECTIONS OF SCHOLARSHIP AND POLICY

Lunetta's early career interests in international science education led to his service as a Fulbright-Hays Senior Scholar in Portugal at a time of great change in schooling, in higher education, and in national development there. In that position, he co-authored papers and three monographs in Portuguese for science educators and science teachers. He also co-taught related workshops with science teachers and teacher educators in Portugal. Subsequently, he served as chair of a Discipline Advisory Committee for the Council for the International Exchange of Scholars [Fulbright]. After retiring from Penn State, he collaborated with colleagues and supported development of science education scholarship in Taiwan under sponsorship of Taiwan's National Science Council. Lunetta's roots in science education and his international activities, scholarly pursuits, and sensitivity to deeply rooted cultural expectations and practices enabled him to assist in policy-related guidance for science education in other countries.

In January 1985, Lunetta accepted a position as Program Director in the Directorate for Science & Engineering Education at the National Science Foundation, where he served almost 2 years. In this role, he oversaw the selection, development, and implementation of innovative funded projects designed to explore and enhance science teacher education. Subsequently, he accepted a faculty appointment in Science Education and Science, Technology, Society at The Pennsylvania State University for 15 years. During that time, he served 5 years as Associate Dean for Research and Graduate

Studies in the College of Education, and for several years as Professor-in-charge of Science Education. At Penn State, Lunetta worked closely with faculty colleagues in the Colleges of Science, Engineering, and Education, and with graduate students in Science Education, to revise and develop introductory science courses for science majors and "core science" courses for students majoring in other fields. Global goals for the enhanced courses focused his work on promoting interest and understanding in science and engineering, scientific inquiry, and scientific understanding. These efforts included development and assessment of curriculum and teaching materials to engage diverse students in solving authentic problems using *inquiry-empowering* technologies. In 2006, four years after his retirement and emeritus recognition, Penn State's College of Education honored Lunetta with its Cotterrill Award for his continuing leadership in the College and the field of Science Education.

Throughout his academic career, Lunetta's leadership in science education could be best described as unwavering and dedicated service to professional associations and public service. He conceptualized the science and educational content for *What's in the News—Space*, a former public television series, and he organized symposia selected by critical reviewers for presentation at meetings of the American Association for the Advancement of Science and other national associations such as the National Association for Research in Science Teaching and the American Education Research Association. Late in his tenure at Penn State and early in retirement, Lunetta served as co-PI on two NSF grants. One grant awarded to Penn State focused on developing leadership in research on learning and teaching in mathematics and science with digital technology tools. The second grant to the NSTA, collaborating with the National Research Council, the Association for Science Teacher Education, the National Association for Research in Science Teaching, the American Association for the Advancement of Science, and other organizations in science and science education, supported the development of the *Research Agenda in Science Education* project. The particular work with his colleagues on the research agenda project led to a sea change at the NSTA. It resulted in inclusion of research on science learning and teaching as one of the four core goals in the NSTA's 2005 strategic plan, the first time that research had ever been included in the Association's goals. Subsequently, NSTA publications and professional development activities actively promoted scholarship to inform science learning, teaching, and curricula.

Lunetta's extensive background, experiences, and scholarly productivity in teacher education led to nationally prominent service. His elections to the Board of Directors for the Association for Science Teacher Education, to the Board of the National Science Teachers Association (twice), and to an Education Panel for the American Association for the Advancement of

Science, signaled recognition of his influence on science education. He was recognized and honored by peers through election as an American Association for the Advancement of Science Fellow, particularly for his leadership in science teacher education. Lunetta received numerous recognitions and awards for teaching, scholarship, and professional leadership in science education and in teacher education. He was named Outstanding Science Educator by the Association for Science Teacher Education, was co-recipient of an OHAUS award for College Science Teaching, and received two STAR awards from the National Science Teachers Association. He also served in principal editorial roles on editorial boards of national and international journals and publications, and continues to be an active, published reviewer of books, papers, software, and DVDs.

CONNECTIONS BEYOND SCIENCE EDUCATION IN RETIREMENT

Lunetta's early life experiences growing up in the northeastern United States initiated lifelong connections to people, issues, and history of the New Hampshire White Mountains. Long-term experiences there helped him develop deep appreciation for the ecosystems and geology of the White Mountains, a strong sense of place in that region, and an understanding of the importance of active environmental stewardship. This understanding, coupled with his lifelong experiences in science education, enabled Lunetta (in retirement) to contribute to conservation and environmental education in the region. His postretirement activities have included service as a volunteer naturalist for the Appalachian Mountain Club, co-chair of a State Park Association Weeks Act Committee for 5 years, and active membership and participation in the Inter-Agency & Inter-Association Weeks Act Centennial Coordinating and Education Committees. His work with others on these committees contributed to the development of public resources that currently include a Legacy Trail supported by a website, print brochure, video, and two DVDs. The DVDs document the 100+ year history of intense environmental degradation of the eastern Appalachian forests and mountain ecosystems with older video footage and slides. They review the intense grassroots citizen action and the political leadership that enabled passage of the contentious Weeks Act of 1911. The DVDs show that the Weeks Act resulted in the establishment and preservation of *all* national forests east of the Mississippi River; the informed management of millions of acres of mountain ecosystems, forests, and watersheds; and the subsequent renewal, productivity, and beauty of the formerly wasted lands. New threats continue to emerge, and the DVDs articulate the importance of continued informed management of these now-treasured resources and relevant political action to assure that these much-loved resources are sustained.

Lunetta's role as a pioneer in science education and his influence on the field continues well into the retirement phase of his career. When asked what he thought his greatest accomplishments were, he modestly stated,

> The energetic professional engagement, teaching, service, and accomplishments of numerous former students has been wonderful to observe. Like a concerned parent, I cannot take any personal credit for what they have accomplished, I can think I may have encouraged some of their work, and I can assume that the role I played in their development did not excessively limit what they were able to do.

A quote that is representative of Vincent Lunetta's philosophy, approach to life, and career as an outstanding educator.

REFERENCES

DeLaeter, J. R., & Lunetta, V. N. (1982). Environmental issues: A responsibility of science teachers. *Australian Science Teachers Journal, 28*(3), 5–10.

Fuhrman, M., Lunetta, V. N., & Novick, S. (1982). Do Secondary School Laboratory Texts Reflect the Goals of 'New' Science Curricula? *Journal of Chemical Education, 49*(7), 563–565.

Hofstein, A., & Lunetta, V. (2004). The laboratory in science education: Foundation for the twenty-first century. *Science Education, 88*(1), 28–54.

Lunetta, V. N., & de la Rosa, L. M. L. (1983). Aspectos ambientales: Recursos importantes para todos los programas de ciencia. *Biologia, 13*(1–4), 49–52.

Lunetta, V. N., & Lederman, N. G. (1998). Reform of science education in Taiwan: Contexts, opportunities, and concerns. *Proceedings of the National Science Council, 8*(3), 121–131.

Lunetta, V. N., & Novick, S. (1982). *Inquiring and problem solving in the physical sciences: A sourcebook.* Dubuque, IA: Kendall/Hunt.

Lunetta, V. N. & Tamir, P. (1979, May). Matching lab activities with teaching goals. *The Science Teacher, 46*(5), 22–24.

Lunetta, V. N., & Tamir, P. (1981, December), An analysis of laboratory activities: Project Physics and PSSC. *School Science and Mathematics, 81*(8), 635–642;

Tamir, P., & Lunetta, V. N. (1978). An analysis of laboratory inquiries in the BSCS yellow version. *The American Biology Teacher*, 40(6), September, 353-357.

Selected Bibliography of Lunetta's Work

DeLaeter, J. R., & Lunetta, V. N. (1982). Environmental issues: A responsibility of science teachers. *Australian Science Teachers Journal, 28*(3), 5–10.

Fuhrman, M., Lunetta, V. N., & Novick, S. (1982). Do secondary school laboratory texts reflect the goals of 'new' science curricula? *Journal of Chemical Education, 49*(7), 563–565.

Hofstein, A., & Lunetta, V. (1982). The role of the laboratory in science teaching: Neglected aspects of research. *Review of Educational Research, 52*(2), 201–217.

Hofstein, A., & Lunetta, V. (2004). The laboratory in science education: Foundation for the twenty-first century. *Science Education, 88*(1), 28–54.

Lunetta, V. N. (1974, September). Computer-based dialogs—A supplement to the physics curriculum. *The Physics Teacher, 12*(6), 355–356.

Lunetta, V. N. (1977, April). Human transactions and classroom management. *Educational Technology, 17*(4), 35–37.

Lunetta, V. (2003). The school science laboratory: Historical perspectives and contexts for contemporary teaching. In B. Fraser & K. Tobin (Eds.), *International handbook of science education* (pp. 249–262). Dordrecht, The Netherlands: Kluwer Academic.

Lunetta, V. N., & Dekkers, J. D. (1982, April–June). Needed: New models for linking the two cultures. *European Journal of Science Education, 4*(2), 149–160.

Lunetta, V. N., & de la Rosa, L. M. L. (1983). Aspectos ambientales: Recursos importantes para todos los programas de ciencia. *Biologia, 13*(1–4), 49–52.

Lunetta, V., & Hofstein, A. (1981). Simulations in science education. *Science Education, 65*(3), 243–252.

Lunetta, V., Hofstein, A., & Clough, M. (2007). Learning and teaching in the school science laboratory: An analysis of research, theory, and practice. In S. Abell & N. Lederman (Eds.), *Handbook of research in science education* (pp. 393–342). Mahwah, NJ: Erlbaum.

Lunetta, V., Hofstein, A., & Giddings, G. (1981). Evaluating science laboratory skills. *The Science Teacher, 48*(1), 22–25.

Lunetta, V. N., & Lederman, N. G. (1998). Reform of science education in Taiwan: Contexts, opportunities, and concerns. *Proceedings of the National Science Council, 8*(3), 121–131.

Lunetta, V. N., & Novick, S. (1982). *Inquiring and problem solving in the physical sciences: A sourcebook.* Dubuque, IA: Kendall/Hunt.

Lunetta, V. N., & Tamir, P. (1981, December). An analysis of laboratory activities: Project Physics and PSSC. *School Science and Mathematics, 81*(8), 635–642.

Lunetta, V. N., & van den Berg, E. (1995) Tailoring science education graduate programs to the needs of science educators in low-income countries. *Science Education, 79*(3), 273–294.

Lunetta, V. N., & Zalewski, L. J. (1974). *Interactive incidents from classroom, school, and community.* Washington, DC: National Science Teacher Association.

Penick, J., & Lunetta, V. (1989). Societal and educational applications of physics concepts. *The Physics Teacher, 27,* 287–290.

Poduska, E., & Lunetta, V. (1984). The role of applied physics in American introductory physics courses. *Physics Education, 19*(5), 241–246.

Tamir, P., & Lunetta, V. N. (1978). An analysis of laboratory inquiries in the BSCS yellow version. *The American Biology Teacher, 40*(6), 353–357.

Tamir, P., & Lunetta, V. (1981). Inquiry-related tasks in high school science laboratory handbooks. *Science Education, 65*(5), 477–484.

Taylor, J. A., Lunetta, V. N., Dana, T. M., & Tasar, M. F. (2002). Bridging science and engineering: An integrated course for non-science majors. *Journal of College Science Teaching, 31*(6), 378–383.

Additional Bibliography

Lee, Y. (2007). Facilitating conceptual change in physics through laboratories and computer simulations: Vince Lunetta's contribution to science education. In X. Liu (Ed.), *Great ideas in science education: Case studies of noted living science educators* (pp. 63–74). Rotterdam, The Netherlands: Sense.

CHAPTER 10

VERNE ROCKCASTLE

Life-Long Inquirer, Teacher, and Innovator

Edward L. Smith
Michigan State University

Jane Heinze-Fry
Museum Institute for Teaching Science, Quincy, MA

Ellie Horwitz
MA Division of Fisheries & Wildlife (retired)

Don Duggan-Haas
The Paleontological Research Institution and its Museum of the Earth & Cayuga Nature Center

Verne Rockcastle was born with an inquiring mind in a small wiry frame to a mother who focused on rearing him and a father who led with physical strength and incredible carpentry skills. It was here that the acorn of Verne Rockcastle sprouted and grew and eventually was carried far enough away to grow and thrive into a mature oak in his own right. This science education pioneer then spread his ideas to and through his students, both

Going Back for Our Future II, pages 165–190

undergraduate and graduate; teachers' workshops; and writings in science leaflets, magazine articles, textbooks, and the web.

As we who knew Rocky for only a small part of his many years of teaching at Cornell University (including nearly two decades after retirement) reflect upon his contributions, we must ask about the lifelong experiences that informed the development of his mind and spirit. His autobiographical *Acorn to Oak* (Rockcastle, 2009) provides much of the background for the following insights.

YOUTH: CITY AND COUNTRY DURING THE GREAT DEPRESSION YEARS

Rocky's resourcefulness, attention to detail, and appreciation of precision were likely influenced by his father, as illustrated in a project they undertook involving crafting their own fly rods from scratch. Their ingenuity and persistence was impressive. Each section of the rods consisted of six lengths of bamboo, each an equilateral triangle in cross section, glued together to make a stick hexagonal in cross-section. The sections of bamboo were placed in 60-degree grooves in a long block of maple where they were shaped with planes, chisels and scrapers to form the tapered elements composing the poles. This was also an early beginning to his lifelong love of fly fishing. (Rockcastle, 2009, p. 12)

As a youngster growing up in the Depression, he built a communications system with fishing line to send notes to his friend who lived across the street, raced cars made from orange crates, built a portable clubhouse with his friends, and worked with his father on shingling roofs in the summer. The natural world was his to explore at his grandparents' farms and in summer months spent in the Adirondacks. He observed and explored birds, fish, mountains, lightning, cattails, and redwings. He took interest in how natural resources were used as raw materials in industries: he observed logging of birch, beech, and maple, and the creation of hardwood bowls, Dixie cup spoons, and tongue depressors at the Oval Wood Dish company. He reflected, "Hour upon hour was spent this way observing everything that had wings, legs, or fins" (Rockcastle, 2009, p. 37). Close observation of mayflies and kingbirds, barn swallows and bitterns, and bass taught him about predator-prey, adaptations, territory, and competition. We make a point of this because we know this style of childhood, while fairly common in that era, is rare in this century (Louv, 2008). The theme of "real-world applications of knowledge" continued to develop through Rocky's life.

UNDERGRAD: OVERCOMING BARRIERS AND FOCUSING ON SCIENCE EDUCATION

Loving the outdoors, but with ambitions beyond being a forest ranger, Rocky chose to do his undergraduate education at the College of Forestry at Syracuse University. He considered majoring in Pulp and Paper Engineering, which would blend science and math, but discovered that he couldn't tell the end of the titrations because of color-blindness. He transferred to Landscape Engineering, and when color-blindness remained a barrier to that field, finally to Liberal Arts with a major in Science Education. He listened to the passing trains and thought of the Doppler effect. He created flash cards to learn the insects. He counted the chirps of the snowy tree cricket, added 40, and determined the degrees Fahrenheit. He would walk home from dates with Madeline, who would later become his wife and lifelong companion, and review the constellations. And in so doing, he found his niche!

He wrote,

> Preparing for science teaching was so rewarding! I had to think of creative and effective ways to get kids to think about and to understand concepts of science. It wasn't enough to learn that certain things happened in physics, chemistry, biology, astronomy, and earth science. If I was going to be a good teacher, I had to find ways to show kids why they happened the way they did. Teaching science was going to be more fun than doing science! As I learned the whys of science, and discovered neat ways of explaining science, I realized that science teaching was what I wanted all along. Science teaching was going to be a great profession. Better than finding out what people (kids) know would be to find out how they know, and especially how to convince both the ignorant and the skeptical. (Rockcastle, 2009, p. 54–55)

He graduated in May 1942, with an AB degree, cum laude, a "certified, but unskilled science teacher" (p. 55).

MILITARY SERVICE: LEARNING AND TEACHING

In July, following graduation, he was inducted into the U.S. Army. He found himself in the Army Air Corps and off to Atlantic City for basic training. Assigned to Lowry Field, Denver, Air Force Armament Unit, he applied and broadened his knowledge of armament mechanics: taking apart and reassembling a .50 caliber machine gun; appreciating the hydraulic power units that rotated and elevated the machine guns on the B-24s. After training at Lowry, many of his platoonmates were assigned to the South Pacific, while Rocky had the good fortune to be assigned to MIT in Boston for meteorology training, "And all because I had done well on an exam and filled out a seemingly unimportant form at the reception center. Thus began a most important part of my academic and professional career" (Rockcastle, 2009, p. 64)

To develop the men's meteorological skills, the Air Force had brought in weather researchers from New Zealand, Iceland, Norway, and other places where weather was spawned. This gave Rocky access to the cutting edge of this rapidly developing international science. As they worked to provide meteorological data for the troops, Rocky realized the need for computers. They weren't far enough developed at that time to find an important place in meteorology, but the handwriting was on the wall. Within a few years, more data had been gathered and interpreted in meteorology than in any other subject.

Later, he and his graduate students would experiment and apply computer technology to education in a variety of ways. After MIT, his geographic sense was broadened by postings to different bases around the country: Seymour, Indiana; Randolph Field, Texas; Albany, Georgia; and Orlando, Florida.

TEACHING IN ROCHESTER

Rocky began his public school teaching career as a substitute science and math teacher in the eighth grade at Madison High School in Rochester, New York. His military experience gave him the confidence he needed to address the challenges of substitute teaching. His pedagogy developed more fully as he taught in the Veterans Unit at Jefferson High School. This assignment offered interesting experiences that illustrate how Rocky was to become a pioneer of "real world" science. He tells the following anecdote:

> One thing I did that seemed to please and totally involve the class was taking the science out-of-doors in a practical, applied way.... One of the guys had a "souped-up" car (of early 30's vintage) and he claimed it could "put out" something like 60 horsepower. We decided to test this, and with the cooperation of the Rochester Police, we cordoned off a city block just off Bloss Street, near the school, chalked a starting line and a finish line on the street, weighed the car, and put it to the test. When we plugged mass and acceleration into our formula, the car produced a huge total of something like 30 horsepower! The students took a more than casual interest in science from then on! (Rockcastle, 2009, pp. 92–93)

Other experiences included analyzing a catcher's throw to second base and investigating the effects of chrome and black cooling fins on air-cooled engines.

Rocky tells about how the old house which he and his wife bought in a Rochester suburb became one of his teachers, foreshadowing his improvising of science equipment in much of his teaching. He writes,

> I learned that if you didn't have the wherewithal to purchase, but had access to some things that could be cut, glued, bent, or drilled, and had both creativity and determination, there really wasn't much of a limit on what you could accomplish. And it was fun being inventive! (Rockcastle, 2009, p. 89)

As he was looking for a permanent teaching position, his principal at Jefferson recommended that he apply for an open position at Brockport State Teachers College and arranged an interview for him. His having been selected as a meteorology instructor in the military at MIT supplemented his short teaching experience in Rochester and helped him get the position.

BROCKPORT STATE TEACHERS COLLEGE

During the Brockport years, Rocky designed and oversaw the building of his family's home, inspired and coached a cross-country team, and taught science to aspiring teachers. He took special interest in evaluating a summer program for kids from lower socioeconomic strata.

Rocky's high expectations for himself were reflected in his teaching of biology at Brockport. He learned what it meant to teach a full schedule. He was determined to become an exceptional college teacher. So when he was offered the opportunity to teach human anatomy and physiology, he jumped at the chance. He ordered the best-known textbooks used by medical students and developed laboratory exercises and lectures. He and the students suffered through that first year but survived and developed an *esprit de corps.* He found himself being asked (and answering) questions that he had never been asked before. When asked some embarrassing questions about sex, he brought in a local physician. His fame was spreading, because no other classes had ever had such a straightforward discussion of such matters.

He was determined to make the laboratory part of his biology courses "real" and so spent many lab periods outdoors. His classes

> investigated a small stream... tore bits of bark from fallen, decaying trees and examined the flattened wood borers beneath the bark.... turned over stones to examine invertebrates underneath... and listened to bird songs and learned the repertoire of some of the most common songsters.... Other classes studied Planaria from prepared slides. My classes visited local ponds to collect live Planaria. (Rockcastle, 2009, p. 100)

While teaching future teachers, he took them into the real world of children. For example, he took a group of his students and a kindergarten class on a field trip to a local swamp where the children cooked bacon and eggs on #10 can stoves fueled with twigs.

Meanwhile, Dr. Victor Schmidt, a PhD from Cornell, had joined the Brockport staff as a geologist. Vic introduced Rocky to workshops for teachers that opened a new avenue of professional experience. Vic had an intuitive understanding of how elementary school children learned and how to prepare teachers for that level of science teaching. Throughout central New York State, Vic and Rocky offered teacher workshops that were both effective and popular.

Rocky soon learned that, while he could continue teaching at Brockport, he could not become a full professor without an advanced degree. He was inspired by attendance at a workshop by Dr. E. L. Palmer from Cornell University, the author of the *Fieldbook of Natural History* (1949). Palmer was a superb field naturalist who did not lecture about natural history but rather immersed his students in it! Rocky resonated with Palmer's style. Rocky had found Cornell to be the intellectual home where he would pursue graduate study for the PhD.

The chair of Rocky's graduate committee was Dr. Eva Gordon, who watched over his research and thesis like a mother hen. He thought the world of her, but was not used to having someone be quite so intimately demanding, and he resented it a bit. Although he may not have appreciated it at the time, this exacting care later become a hallmark of his own work with graduate students.

Rocky's writing career began as a doctoral student. Dr. Gordon, who had gotten behind schedule, asked if he would be interested in writing one issue of the *Cornell Rural School Leaflet.* The Leaflet was a 32-page quarterly for rural schools of the state. They covered a variety of science topics, especially natural history ones. Rocky agreed and wrote one on *Air Laboratories* (Gordon & Rockcastle, 1953). It was about weather near the ground, how to study it, and how to measure it. It seemed to be at the right level for children and teachers, and he found it fun to do. The "author bug" had bitten him (Rockcastle, 2009, p. 119).

Rocky then looked about for other writing opportunities, especially ones that paid a little, and found *The Instructor,* a magazine for elementary teachers. He achieved success with an article about a local elementary school in which he wrote about geology, with pictures of puddles and alluvial fans. Rocky wrote two more grade-level articles on science and then was awarded a 3-year contract by the magazine to do more. He now felt he was on his way as an author of children's science publications and had the benefit of being paid to do it.

Commuting to Cornell, Rocky studied microclimates—an extension of some of the research at MIT, which then became the substance of his PhD research. He was fascinated by the climate within a few centimeters of the soil surface, especially as it affected four kinds of animals living in that lowest layer of air—grasshoppers, salamanders, voles (meadow mice), and even woodchucks. What was their climate like? What were the temperature and humidity extremes? What precautions or peculiarities of habit enabled them to minimize the effects of temperature extremes? His degree was conferred in August 1955. After the retirement of Eva Gordon in 1956, Rocky found himself at Cornell University as a member of the faculty for the next 30 years of his career.

CREATING A NICHE AT CORNELL

Arriving as a faculty member at Cornell, Verne continued to write the *Cornell Rural School Leaflet* (which subsequently became the *Cornell Science Leaflet*) and taught an elementary school science course for teachers. From his Brockport role as track coach, Rocky continued his support of track and developed a unique role as an advisor to the Cornell track team. But Cornell brought new responsibilities and opportunities. He had a new and different responsibility—graduate students pursuing MS and PhD degrees. He found this quite different from working with undergraduates. He was to oversee graduate research and the writing of theses. "I encouraged creativity, and engaging in new and creative research…working with teachers and students in public schools. Go where the action is! If it doesn't exist, design it and determine its efficacy!" (Rockcastle, 2009, p. 126).

Rocky went on to advise 20 PhD students and 43 MS students. He notes, "Only twice in my twenty-five years of directing graduate students did I have students who didn't complete what they started, who didn't find graduate research as challenging and fulfilling as I did" (Rockcastle, 2009, p. 126).

While advising and mentoring graduate students was a new role, Rocky soon discovered an even greater difference between the college at Brockport and Cornell University, one he came to fully embrace. He was expected to generate new programs without benefit of colleagues. Initially he felt totally alone. His science education colleagues each had their own "orbit."

> *But what was mine?* It took me essentially three years to find my own niche. That was the essential difference between college and university! It was frightening, but once I realized and accepted that, it was collegiate heaven! I was my own boss, and I was encouraged to be creative! Cornell was indeed the right choice! (Rockcastle, 2009, p. 126)

Rocky's basic load included teaching elementary science education and mentoring graduate students, in addition to writing the *Cornell Science Leaflets.* In creating his own unique niche, he expanded his author role by writing articles for *Science and Children,* co-authoring the book *Teaching Science with Everyday Things* with Victor Schmidt (1968), and authoring an elementary science textbook series for the Addison Wesley publishing company (Rockcastle, Salamon, Schmidt, & McKnight, 1970). Rocky expanded his work with elementary teachers beyond the university conducting many workshops in the region, at NSTA conferences and in conjunction with his text series. In his unique niche, he interwove these activities so that each informed and enriched the others. Phenomena encountered in his science courses or in his explorations inspired ideas for investigations for children. Ideas for investigations for children found their way into his courses and workshops for teachers. His graduate students were mentored in designing and conducting workshops for teachers and in writing projects such as the *Cornell Science Leaflets.*

PIAGET REDISCOVERED

A unique opportunity arose during Rocky's first sabbatical from Cornell when, while in Europe, he was asked to meet with Jean Piaget to try to arrange for him to come to the United States to present lectures or participate in a conference. With his marginally functional French, Rocky introduced himself, as "Professor Rochechateau from Cornell," and Piaget's agreement to participate was obtained, assuming that the proper arrangements could be made. Arrangements were made with support from the National Science Foundation and the U.S. Office of Education, and a two-coast conference was planned and conducted at Cornell and the University of California at Berkley. Rocky was the overall head of the conference. Although translators for both French-English and English-French were obtained, they were little used. Eleanor Duckworth, one of the conference participants, had studied with Piaget and knew both him and his theories well. She agreed to translate Piaget's lectures and was important to the success of the Cornell conference.

The impacts of the conference on both the field and a personal level are reflected in comments by Ted Bredderman.

> By the time I entered the Science Education program at Cornell, Verne Rockcastle along with Dick Ripple had become well known nationally for inviting Jean Piaget to a conference in the United States, 1964. Important papers at the Conference on *Cognitive Studies and Curriculum Development at Cornell* were published under the title of "*Piaget Rediscovered*" [Ripple & Rockcastle, 1964]. At this same time, NSF involvement in elementary science curriculum

> development was just getting started and the Ripple-Rockcastle publication became a major influence for some of the NSF supported projects, for example *Elementary Science Study* [ESS] and *Science Curriculum Improvement Study* [SCIS], and others. The conference and publication that it generated were the start of a line of work which led to a number of doctoral and independent research studies of which my own studies on controlling variables were examples [e.g., Bredderman, 1974]. In the decades following the publication of *Piaget Rediscovered*, science education scholars made significant contributions to the literature on theoretical issues related to cognitive development, and... how the learning of science can give us insight into cognitive development.... Clearly this initiative by Ripple and Rockcastle was a major turning point in science education in America and across the world. Thirty and forty years later some of my international doctoral students in science education came to this country wanting to pursue studies based on Piaget's work. (T. Bredderman, personal communication, August 30, 2009)

Rocky's interest in Piaget and Bredderman's research proposal inspired one of this chapter's authors (Edward Smith) to serve as Ted's unpaid research assistant. That experience in turn led to Dr. Smith's career-long appreciation of the importance of taking into account children's operational capabilities when designing and developing curriculum.

Although less prominent in the conference, Piaget's earlier work in exploring children's conceptions emerged as his work was "rediscovered." This fascinating work inspired a huge body of research leading to the awareness, at least among researchers, of the strongly held ideas about the world around them (often referred to as "misconceptions") that learners bring to the classroom (e.g., Driver, Squires, Rushworth, & Wood-Robinson, 1994). An American Association for the Advancement of Science's website lists a bibliography of more than 270 references on "Misconceptions research."[1]

A UNIQUE PHILOSOPHY

Rocky had a unique commitment to engaging learners in seeing and learning from the wonders of objects and events in the world around us. This reflected his own lifelong fascination with how and why things work and his infectious enthusiasm for pursuing these questions. His philosophy of science and education is well represented in the following comment from Ted Bredderman, who wrote,

> Verne's most fundamental impact on me personally was in the area of philosophy of learning and teaching in science. Verne seldom if ever expounded on philosophy but he was clearly grounded in a strong set of beliefs about science and education. He walked the walk almost exclusively and seldom talked the talk. I suspect that one of his closest attempts at expressing general claims

> about learning and teaching is found in the opening remarks to his book "Teaching Science with Everyday Things" by Schmidt and Rockcastle (1968) in which he wrote: *The purpose of these questions is to encourage pupils to observe and think carefully... No answers are given,* because observation and thought—*not the book or the teacher*—should be the source of the answers.
>
> I believe that this faith in the teacher's ability to present questions and experiences which promote student observation, thought, and in turn student learning, are his most basic foundation for teaching science to students at all levels and for preparing educators for leadership roles in the field of science education.
>
> At first glance, activity-based teaching appears to be too simple an approach to be considered significant, but in the end, it is at the core of getting nature to reveal its secrets to students. If nature's secrets were easily observed, there would be no need for science education, science teachers, and science activities. But, nature seems to hide how it works so a good teacher contrives ways of having nature expose its mysteries. A good activity presents mysteries, which can be explored at various levels. Verne was a master in this respect, a master activity engineer. One can see the influence of Dewey, Piaget, Vygotsky, Gagne, Karplus and others in his approach, but it is not hard to recognize the package when you see it as clearly "Rockcastle." For sure, his activities inspired my curiosity, and my thinking about nature, teaching, and about learning. And more often than I wish to admit, made me an imitator of the master. (T. Bredderman, personal communication, August 30, 2014)

Jane Heinze-Fry, a doctoral student from the 1980s, phrased it differently saying,

> He [Rocky] modeled a wonder at the world and its natural phenomena and explanations of how the world worked. And he encouraged the development of this curiosity in others through his modeling and guiding questions. He would invent ways to explore complex phenomena and concepts in a concrete way. He applied his understanding of science and engineering in his daily lifestyle. He was a constructivist, both literally and figuratively. Explaining phenomena in the real world such that we could grasp the meaning: "How do things work?" he would ask. "Now look more closely. Tell me more about how it works." What hoops did we jump through as we learned in this manner?
>
> - Outside, we modeled the solar system with our bodies. We walked appropriately to demonstrate an eclipse.
> - We jumped on pieces of appropriately spaced masking tape on the floor and clapped our hands to get both a physical and auditory understanding of the concept of acceleration.
> - At a table, we moved the parts of a car's transmission to investigate how gears change.

- We built radios to understand how this form of communication works.
- We used a thermocouple to measure temperature and create a topographic map of the varying external temperatures of the human body.
- We learned how cameras worked, developed film in the dark room, made and mounted our own prints, and participated in photography contests. (J. Heinze-Fry personal communication, August 24, 2014)

Doing "hands-on, minds-on" science has been advocated by a number of science educators. Rocky's use of "everyday things" represented more realistic inquiry than many others' approaches. Rather than laboratory equipment or contrived materials, his investigations drew on the world as experienced in and out of the classroom, often using ingenious homemade instruments. His questions often posed practical or engineering-oriented problems rather than straight description or explanation questions. The questions would relate to a big idea and require further investigation. He would include multiple examples for a concept. His emphasis on everyday things and homemade instruments made his approach less expensive than many other approaches.

CURRICULUM/PUBLICATIONS

Rocky's contribution also appears in the form of the voluminous collections of activities that dominated his writings and his face-to-face teaching.

The Cornell Science Leaflets (Originally the Cornell Rural School Leaflets)

For several decades, Rocky guided science education through the *Cornell Science Leaflets.* He wrote and published four issues annually. He was not the first author of the Leaflet series. But the direction of the leaflets changed under his leadership. Instead of just nature study, the leaflets took on all aspects of science. They also changed from being didactic to being inquiry oriented with numerous hands-on experiences for teachers to use with their students. From his dissertation research, he produced an issue on "Little Climates" (1961). Former student Bonnie Barr-Larkin wrote, "In an elementary classroom recently, a teacher pulled out a *Cornell Science Leaflet* and commented, 'This is the best resource on the topic of sound that I have seen'" (Personal communication, August 20, 2014). Collections of the investigations were republished with permission by agencies such as the North Dakota Department of Public Instruction (Rockcastle, 1963).

Addison-Wesley Elementary Science Series: Space, Time Energy and Matter

As a co-author of the STEM elementary science text series (Rockcastle et al., 1970), Rocky attempted to be true to an activity-based approach in a textbook format; not an easy task, especially when national forces were encouraging the switch away from textbooks and the NSF was promoting the alphabet programs it had funded—SAPA, ESS, SCIS, and such. Of course, the reality was that elementary teachers were *not* abandoning texts or adopting activity-based science in great numbers.

Initially, Rocky was a junior member of the Addison-Wesley author team. However, the senior authors' contracts were canceled when they delivered several outlines and general plans but no copy-ready lessons for examination by teachers and curriculum specialists. After the other junior author was hospitalized, Rocky was made lead author. He assembled a team of writers and the series was produced.

> Writing a textbook series was perhaps the biggest professional exploit of my life. It was more of a chore, more of a commitment and yet had more professional return than any other educational endeavor of my life. Before immersing myself in this great commitment, I had no idea, really, of the magnitude of the project. The time spent in assembling and outlining, then "fleshing out" concepts; the integrating of the various topics across grade levels; the classroom trials and poring over statistical returns and teacher comments, the testing of individual lessons in classrooms, both urban and rural, and the "selling" of the program to school systems—it was a huge, and at times, a daunting endeavor. But it also had its elegant returns. (Rockcastle, 2009, p. 191)

The first books of the series to be published were in French while a second version was in Spanish. Rocky would not settle for a simple translation. He had to collect illustrations representative of each area. For example, he spent hours photographing and tape-recording crabs on the beaches, coqui (a tropical frog of the rainforest), and many other specialties of lower latitudes and the rainforest.

According to Bonnie Barr-Larkin,

> The activities...went beyond the individual skills to true inquiry, e.g.,
>
> - How might you make an electromagnet stronger?
> - How could you find out if boys have larger lung capacities than girls? Athletes vs. non-athletes? Older vs. younger?
> - How big a mirror would you need to see from the top of your head to the tip of your toes?

- Using what you now know about static electricity, show why a lightning rod is used to protect a building during a lightning storm. (Personal communication, August 20, 2014)

Rocky's improvised science instruments were legend: an equal arm balance for measuring mass, an inclinometer for determining the height of a building or altitude of a celestial object, an incubator for quail eggs, and a homemade magnetizer among them. His inexpensive instruments opened the way to investigating for many with limited resources. But the enthusiasm that Rocky inspired was the major door-opener.

PROFESSIONAL DEVELOPMENT FOR TEACHERS AND OTHER EDUCATORS

Rocky was a master of workshops for teachers. His knowledge, hands-on problem-solving approach to science, and his enthusiastic teaching manner made him a favorite with educators at all levels. His workshops at NSTA conventions were among the best attended of the convention. Through hundreds of workshops he impacted thousands of teachers all across America and in Europe and Israel.

On occasion, his workshop audiences included textbook review committees or classes of teachers at other institutions. In his autobiography, Rocky relates his experience in Texas:

> In each school region of Texas, from El Paso to Beaumont, from Amarillo to Brownsville, the state school board demanded a presentation... for local school boards, teaching staff, school administrators, and specialty groups.... [At] a small teacher education college.... I prepared a one-day program of science for elementary teachers, held in their gym... The presentation was a "smash" and as a result, I was asked to present a one-week summer course... for teachers.... It was, as with all my classes, a hands-on, experiential course, and it was an immediate success. I was asked to teach a similar course the next summer, and the next,... for 12 summers. (Rockcastle, 2009, p. 194)

Retirement did not slow down his efforts in either writing or in-service. For years, he would travel to Alaska to conduct a 2-week in-service for the Alaska Science Consortium.

IMPACT THROUGH MENTORING OF GRADUATE STUDENTS

As evidenced by the responses to our inquiries of former students for this chapter, Rocky had profound impacts on many of those whom he mentored. Perhaps his most significant influence was as a role model.

A Model for Professorship

Ted Bredderman writes,

> An area in which Verne influenced me, was in the role of a professor. After receiving my PhD, I became an Assistant Professor of Education at SUNY at Stony Brook. What I knew about being a professor I had acquired through observing the professors I had encountered at Cornell over a period of about eight years. There are no "professor training" courses to introduce you to the process. As one of my professors, my advisor, chair of my dissertation committee, and co-supervisor of my graduate assistantship work over a period of three years, I can say that Verne was my "professor training." He set the bar very high, giving of his time, and expecting much. His attitude and effort made you want not to disappoint him. I can only hope that my professorial behavior had the same positive effect on the many students I taught and supervised over the years. (personal communication, August 30, 2014)

Bonnie Barr-Larkin reflects,

> Verne would take his graduate students to meet and interact with noted science educators. I remember our visit with E. Laurence Palmer. I have his trademark rabbit track signature in my copy of Palmer's *Fieldbook of Natural History* (1949). Verne encouraged his graduate students to attend the NSTA conventions and frequently arranged for transportation.... His graduate students shared office space in what was known as the "grad room." Verne's office was right around the corner. When he had an exciting idea that just had to be shared he would pop around the corner and say something like, "What do you think of this?" With Verne as a mentor there was never.... "Pride of Position." His graduate students were treated as equals in a quest for developing excellence in science education. (Personal communication, August 20, 2014)

Jeanette Bakke writes that Rocky was a stickler with rules, and yet he was also compassionate. "In the spring I arrived in his office totally overloaded. He helped me prioritize what needed to be done in all sectors and then supported me along the way as I worked the plan.

Michael Cohen observed that, "Verne Rockcastle was a unique role model and mentor who taught by example, with real objects and experiences, while treating all with respectful challenges. He also was as interested in our personal growth and development as well as intellectual" (Personal communication, July 3, 2014).

Jane Heinze-Fry notes,

> As a teacher/educator of graduate students, he wanted you to be the best you could be: to have research under your belt that would make you unique, that

you could go out into the world and offer in a unique way. And meanwhile, he would constructively criticize any sloppiness in your use of technology, your attire or your use of language, any fuzziness in your thinking and explanations of scientific phenomena. (Personal communication, August 24, 2014)

Don Duggan-Haas came to know Verne long after he had "retired." Don writes,

I began as a visiting faculty member in Cornell's Education Department, working primarily with future science teachers. Though Verne had retired many years before, he was still in his office nearly every day, and teaching his course on the science of everyday things regularly. There were no teaching labs dedicated to (or easily available for) science teacher preparation, so I had to figure out how to engage my science teaching methods students as best I could in regular university classrooms.

Fortunately I met Verne within days of starting at Cornell, and expressed my frustration about the lack of supplies. He maintained a great stock of things, in his own stockroom, hidden away. He told me with pride about how he long ago procured the room. He noticed that it appeared to be unused for a couple of semesters in a row, and tested that by setting up a sort of booby trap—just to make sure. He attached a string to the doorknob with the other end connected to something—I no longer recall exactly what—that would be obvious if it was disturbed. It went undisturbed long enough that he simply took over the little store room and made it his own. Before too long, he'd changed the locks so he alone had access—until he allowed me into the realm. (I never got my own key, though).

He was delighted to have a colleague actually interested in teaching science with things besides text and discussion. He shared not only his physical resources with me, but also many of the activities already mentioned in this chapter, and himself. He served as a guest speaker in more than one of my classes, as did Bill Nye, the Science Guy. Verne was the more engaging and appreciated speaker of the two. He was then over 80 years old and had the future teachers, mostly in their early 20s, both amazed and inspired. (Personal communication, August 20, 2014)

A Midwife for Writers

Rocky was himself a fine and prolific writer. But his contributions were multiplied through his mentoring of other writers. As a grad student under Verne, Bonnie Barr-Larkin was "blessed to be exposed to his theories of education" which she has advocated during her own 43 years of teaching.

As rich and rewarding as my classroom studies with Verne were, it was the enrichment activities in science education outside of the classroom that put the

> polish on my professional education. At a very busy time in his career, Verne asked me, a first year grad student, if I might be interested in rewriting two Golden Books in science which he had written several years earlier. I eagerly agreed and that was the start of my book-writing career (14). (Personal communication, August 20, 2014)

Jeanette Bakke wrote,

> Rocky invited me to be his TA and to write two *Cornell Science Leaflets.* He was a rigorous editor and paid careful attention to the accuracy of details as well as graceful English expression. This was an excellent training opportunity for later writing. I still look at the *Pond Life* (Rockcastle & Bakke, 1967) and *Electric Charges and Circuits* (Bakke & Rockcastle, 1967) leaflets and smile. (Personal communication, July 14, 2014)

Ellie Horwitz arrived at Cornell in 1967. A former science teacher, she was immediately captivated by Dr. Richard Fischer and Verne Rockcastle's approach of immersion in science.

> We were not reading about things or being told about them, we were actually doing them. Field trips became the focus of my life. I became a graduate student of Dick Fisher's but had a strong appreciation for the physical sciences and the way Rocky presented them.
>
> One day quite unexpectedly, Rocky offered me the exciting opportunity to work with him on one of the *Cornell Science Leaflets.* It was a time when there was great administrative concern over having animals in the classroom because the animals might be poorly maintained, but more critically because of potential human health issues such as turtles carrying *Salmonella.* This was truly discouraging to those of us who had enjoyed the care and educational benefits of animals in the classroom. Always positively inclined, Rocky pointed out that regardless of the mandates of the NY State Board of Regents, there would always be live animals in the classroom; 24 or more "critters" who might think it fun to conduct experiments on themselves—and so was born the leaflet entitled *Things to Do With You* [Rockcastle, Horwitz, & Harkenrider, 1968]. Production of this leaflet was a joy and when this publication was completed he allowed me to write "Sleight of Light" [Horwitz, 1968]. Rocky, as an editor, provided a clear direction and a light touch—impressive editing skills that have been key throughout my subsequent career in Fisheries & Wildlife. (Personal communication, September 13, 2014)

Chuck Roth also cites Rocky's engaging him in writing projects as a major source of impact.

> My time at Cornell was greatly enriched by my time with Rocky, although he was not my primary advisor, which task fell to Dick Fischer. Much of my time

> spent with Rocky related to the *Cornell Science leaflets.* Originally I just helped block out the scope of an upcoming leaflet with him and consider appropriate illustrations. He was an absolute stickler for accuracy and thoroughly tested any activities or experiments that were to be included. When I got involved he was writing all the leaflets himself and I was simply assisting. In my second year, I was shocked when he let me write and illustrate my own leaflet on food chains and webs. For me, it was like winning a Pulitzer Prize, because Rocky, himself, set such high standards for the Leaflets. (Personal communication, October 3, 2014)

Opening the World Around Us to Others

Whether teaching a college course, leading a teacher workshop, writing up science investigations or doing everyday things like going through an airport or riding up a ski lift, Rocky found ways to infect others with wonder and excitement when viewing the real world around us.

Michael Cohen was impressed when he accompanied Rocky to Hampton Institute for a short workshop to help the university faculty expand and enliven their science offerings.

> We flew on a four engine Constellation and I was shocked as he explained that the plane could glide to a landing if the engines went out—not what I wanted to hear. Later sitting in the airport waiting for our next flight we sat watching the planes take off and he was busy trying to calculate the angles at which the planes lifted off and what amount of energy this required. I was amazed that he was always looking at the world around him and just thinking or contemplating things. (Personal communication, July 3, 2014)

On another flight with Rocky, one of the author's (Edward Smith) eyes were opened to a new phenomenon when Rocky pointed out a spot of light against the background of the earth below. The spot seemed to move along with the plane. Rocky referred to it as "heiligenschein," the German word for "halo." Fascinated, Edward watched the phenomenon for much of the trip as they discussed its origin. He has often looked for it since, but has only rarely seen it.

One of the many ways Rocky found to open others to that wondrous world he saw was through the Cornell Adult University (CAU). The extent of the travels and breadth of experiences Rocky had in making local contacts, discovering the sites, and leading trips is breathtaking. These trips led to locations in the Finger Lakes region around Cornell, the Adirondack mountain region of New York State, and many more far-flung locales, including the Galapagos Islands and especially Alaska.

One year, there was a CAU tour of three valleys in California, including Yosemite and Death Valley. CAU also offered a week in the Sonoran Desert near Tucson along with a tour of the Grand Canyon. Rocky engaged in both. Those fortunate enough to accompany Rocky on these trips experienced roadrunners, many kinds of lizards, saguaro cacti in bloom, and even an occasional rattlesnake. One such specimen was held up for all to see after being placed in a pillowcase and refrigerated overnight.

In Rocky's view, his CAU trips to Alaska "were among the 'mostest' and best." His connection to Alaska began with an instructor from the University of Fairbanks who decided to take Rocky's course in science for teachers to get some new ideas for the topics in his own course at the University of Alaska in Fairbanks (UAF). The two became friends, and Rocky was invited to come to Fairbanks to teach a week of classes. In addition to classes at UAF, Rocky and his colleague traveled by single-engine airplane to a remote village where they found that the physics teacher "didn't show up for four days. He was off with family and friends on a whaling expedition . . . to stockpile the community larder with whale meat" (Rockcastle, 2009, p. 208).

Rocky worked with the teacher to instill some creative science in the program, and among the activities was one where they measured the air pressure inside the tire of a plane parked on the only runway. They measured the area of the plane's tires on the runway and calculated the weight of the plane. Some 2 years later Rocky encountered a boy in the cafeteria at UAF who said, "Dr. Rockcastle! I was one of the kids who weighed that plane in Kipnuk with a ruler and a tire gauge! That was neat!" In reporting this, Rocky wrote, "I knew, then, that my kind of science was the right kind!" (Rockcastle, 2009, p. 209)

Rocky returned to Alaska for much of another summer, visiting native villages, greeting teachers, and learning about the limits and the needs of Alaskan children. He applied for and was granted a summer of teaching science in Fairbanks, a week in Anchorage, and a week in Auke Bay. Through his developing network of contacts, he had many opportunities to learn, to teach and, yes, to fish and observe wildlife, including a bald eagle he nicknamed "Charlie," who appreciated the innards set out for him from the salmon Rocky caught.

For about 10 years Rocky made annual trips to Alaska. These trips provided the foundation for his nine subsequent CAU trips. Rocky's personal contacts, and his knack for establishing new ones, enriched the trips by opening doors to local sites, such as remnants of the gold mining era, wooly mammoth tusks, and colorful establishments for accommodation. The hit of each trip was the scenic splendor of Denali. It never failed to come up to expectations.

PERSONAL PERSPECTIVES/QUALITIES

Several contributors highlight the personal qualities Rocky modeled and that they hoped to emulate. In Chuck Roth's words, "Rocky's devotion to physical health was legendary and he truly treated his body as if it were a temple." Ted Bredderman remembers being impressed with Rocky's lifestyle.

> Rocky was a model of energy, devotion to his work and family. He was always straight-forward in expressing his beliefs about life and personal behavior but never condescending. He treated teachers and doctoral students as equals and was considered a personal friend by many. He shared his talents and skills, but especially his discipline, attention to detail and focus. These qualities were evident in his personal life, his love of photography, his personal science experimentation, his nature studies, his skiing, and his running. As a senior citizen, his energetic life style continued. In addition to still enjoying running and skiing, he had turned teaching from a vocation into an avocation. Well after his retirement as a Cornell professor, I invited him to do some workshops with teachers in the Albany, New York area who were incorporating invention and design into their science and technology classes. He always stole the show with his clever activities, thoughtful ideas, and good humor. When people of his age were enjoying some time off, he seemed to be as sharp as ever and to be invigorated by the act of teaching. They say "good teachers don't retire they just lose their class." He did retire but still brought the next generation under his spell. (Personal communication, August 30, 2014)

Don Duggan-Haas frequently sought Verne out to borrow from his stock of materials.

> One winter morning, I stuck my head in his office, and he wasn't there. I caught him later in the day and mentioned it to him. He said, "Don, look at my pants." He was wearing ski pants, and said that he'd been to Labrador (a nearby ski resort) and had done three runs. I responded that I'd like to be like him when I'm 83, but went on to note that it was pretty unlikely as I wasn't much like that as a chubby 38 year-old. (Personal communication, September 23, 2014)

Rocky integrated his life at so many levels and he reached for excellence in whatever he pursued. He applied his understanding of science and engineering principles at home and in his work in science education. He instructed to the whole person, not just the mind and he knew the backgrounds from which his students came. As he challenged the students' minds, his interests were concentrated, but not siloed in science education. His mind reached and engaged across the disciplines. For instance, he loved to run throughout his life so he developed into a coach of cross-country. He often invited Dr. William Harlow from the College of Forestry at Syracuse University to campus. Dr. Harlow shared his skill of creating a whistle from a cutting of a branch from a moosewood tree. In later years with a twinkle in the eye, Rocky would create a whistle and play Cornell's alma mater: *Far Above Cayuga's Waters.*

Ellie Horwitz wrote,

> What impressed me most was Rocky's allocation of time and effort in his personal life. During deer season he began each day in the woods "patterning his deer." He explained that the challenge was to watch the deer, become familiar with its daily routine but never to shoot it before the end of the season as that would be a waste of the rest of the deer season. As the season drew to a close we would place bets on whether he would get his deer or not. During the years I was at Cornell, Rocky hunted every day of deer season, carried a full teaching schedule, always had time to chat with students, and always took his deer on the last day of deer season—a feat I never fully appreciated until, more than a decade later and employed by a Fish and Wildlife agency, I began to hunt. Today, more than ever, I stand in awe of his hunting/focusing prowess.
>
> Whatever he did, teaching, hunting, dancing (he and Madeleine were excellent ballroom dancers!) he was totally in it... and that may well be the most important lesson he conveyed to all of us who were lucky enough to have crossed his path—to be in the moment 100%. (Personal communication, September 13, 2014)

A large thread through Rocky's teaching and way was his generosity. Jeanette Bakke recalls that "Rocky was generous with his time, insights, sense of humor, and willingness to work long hours at things that are meaningful. His expectations of others were tied to his high expectations of himself" (Personal communication, July 14, 2014)

For Ted Bredderman,

> Verne was a professor and a friend who went well beyond the call of duty to help his students in their career development and in their personal lives. I recall the student parties at his house in which Madeline and Verne entertained dozens of student at a time making sure new students felt welcome at Cornell and got to know each other.
>
> He was always watching out for professional opportunities for his students. On one occasion, he received a letter from Bob Karplus at Berkeley asking him to identify any student who might be able to join the staff at the newly formed *Science Curriculum Improvement Study* at Berkeley for the summer, 1965. Verne immediately wrote back suggesting that I might be available. I eventually did join SCIS and this contact led to many other contacts and opportunities in my career. Bob Karplus told me later that Verne was the first of only a few who responded to his letter. For this, I am grateful. (Personal communication, August 30, 2014)

Jim Chandler had only one class with Verne but found that Rocky had a profound impact on his passion for life and his science and environmental education teaching.

> He was so full of energy, always illuminated the details of science and the world around me. He taught me the importance of not taking second hand information at face value, and to be an active life-long learner. He also taught me to push my limits of teaching and find simple ways to make science understandable. He made the world of science education come alive, and encouraged me to do the same for my students. He was a proponent of DIY (Do It Yourself!) and being a "maker" way before it was fashionable. In his class we built balances out of wood scraps and pieces of a broken mirror, microscopes from flashlight bulbs and crystal radios around toilet paper rolls.
>
> I was very impressed with his story of visiting an education minister in a tropical country, I believe it was Costa Rica. On the taxi ride to his hotel he noticed the different species of tree frogs in different zones of the forest. Instead of watching TV in his hotel, he rented a car that night and went to the top of the mountain road. He put the car in neutral and coasted down the hill while tape recording the songs of the tree frogs and other tropical forest life. By coasting down the mountain he could use the acceleration of gravity to correlate the zones occupied by different species with the elevation on the mountain. The next day he presented the tape to the local educator as a thank you gift.
>
> I have taken this zest and curiosity about the world around me as a way of life. By his example, I have raised aquatic eggs to find out who the parent organism was and turned a light pole in my school parking lot into a sundial that covered nearly the whole parking lot. Learn by personal experience and careful study and observation was Rocky's credo and it became mine. Critical thinking is very important in my teaching along with an openness to new ideas based on the evidence presented. I also learned the importance of small details that can have a profound impact on how a particular system works.
>
> Overall I will always be thankful for the mentoring I received from Verne Rockcastle which has greatly added to my zest of life, focused my teaching on the scientific process, and made me a much better teacher and life-long learner. Even after thirty-nine years of science and environmental education teaching, in a wide variety settings, I will always have a special place in my heart and spirit for Verne Rockcastle. (Personal communication, August 24, 2014)

When Dick McLeod first met Rocky, he felt that Rocky would have little that would challenge him. McLeod had bachelor's and master's degrees in physics and made physics his major at Cornell. Rocky quickly changed McLeod's opinion.

> One of the first class challenges was the pinhole camera. Now every physics student has studied the attributes of the pinhole camera. What did I have to

> learn? Although I knew lots of theory about pinhole cameras, I had NEVER actually used or made one. Rocky challenged the students to make one and take a picture of the building across from us. We all did—using whatever box we had.
>
> As expected (by me), the picture came out fine. I had calculated the size of the hole, the distance to the film, and, using the film speed, decided on an exposure time. The picture was nearly perfect. Several other students also had success. But then, Rocky asked us to predict what kind of picture we would get if we were to take the picture while classes were changing and lots of students were walking around. We all agreed that we would see the campus with lots of blurred figures in it.
>
> We were surprised to find that the picture of the building was perfect, but there were no people in the picture—not even blurred images. Rocky then asked us to explain our results. My background came to the rescue and I could explain what I saw, but felt chagrined that I did not think about exposure time in this picture. No individual was stationary long enough to appear in the picture. Rocky caused me to think in a way that was different than any of my previous professors.... Rocky has this unique way of presenting science questions that seem, at first, obvious, but the results are challenging. Few physics teachers do that.... What a way to teach science! (Personal communication, September 24, 2014)

Carol Hardy was one of Rocky's PhD students during the years 1976–1981. The thing she appreciated about Rocky was "his incredible fund of knowledge, particularly with regard to publishing, and his flexibility in helping me design my program." She writes,

> My situation was unusual, I was forty years old and at that time was teaching introductory biology at Cornell University and had only my M.A.T. Degree. I was also designing and writing a teacher's guide and study guide to accompany William T. Keeton's college textbook. My mentor, Dr. Keeton, and I decided that if I were to continue teaching at Cornell, I would have to get a PhD. How to do this was the problem—I was a single parent with two young children to raise and support. Gone were the days when I could disappear into the lab and work till all hours, days on end. At that time I was doing research on student study guides, which at that time were useless, as were the teacher's guides. Dr. Keeton suggested that perhaps I could put all the work I was doing developing the guides into a degree in education.... he told me, however, that there was only one person in education that he thought I should work with: Verne Rockcastle. So, off I went to see him, and he could immediately see some great possibilities, and that is how I became Rocky's graduate student.
>
> For five years he [Rocky] mentored me and got me through some tough times. His flexibility and understanding of my situation was amazing. In 1978–79 I had to take the year off to teach Dr. Keeton's large introductory biology course... and then again filled in teaching non majors' biology. In between

times and in the alternate years I worked at taking courses and developing my formative evaluation project on the study guides for my doctorate.

Without Rocky's expertise and understanding I couldn't have done it. My emphasis during this period was to prepare myself to be able to teach introductory biology at the college level, and with that in mind I took courses not only in education, but also in all the various biological disciplines: ecology, neurobiology and behavior, plant anatomy, entomology, etc. Looking back I am really not sure how I did it, but Rocky remained my staunch ally through it all. I remember I was just finishing up my course work and had taken a particularly nasty entomology lab practical when I decided I had had it. I went directly from the practical over to see Rocky and announced that I was never going to take another prelim or lab practical or write another paper at Cornell, that I was done with it. Rocky just laughed and said, "I wondered when you would decide this—let's get this dissertation done." And so we did. (Personal communication, September 3, 2014)

WRAPPING UP THE LEGACY

In later years, writes Jane Heinze-Fry,

I would visit Rocky in his retirement home with my family. He gave my daughter a smoothly rounded rock that he had dove for in one of the potholes at Van Buskirk's Glen when he went on a woods walk with E. L. Palmer when he first came to Cornell. This was part of the mystique of Cornell, in this case the passing on of the "Philosopher's Stone." My daughter went on to become a geology major at U. Mass, Amherst, where she had the privilege of co-leading the geology club and giving a brief speech at the Natural Sciences graduation ceremony. I met another young woman this past year at the National Association of Environmental Educators conference, who is a daughter of one of Rocky's Cornell Adult University students. She told me of how her mother always told her about her experiences with Rocky. This young woman is currently a graduate student studying resilience of ecosystems.

Part of the story of a pioneer species is how it prepares the ground for the following succession. To paraphrase Rocky, "It is our responsibility to raise the next generation to be better thinkers than we ourselves. Essentially, if we only do a brain dump, that's not enough. We should raise the next generation to surpass us." There was a bit of "Dumbledore" in Rocky. (Personal communication, August 24, 2014)

Growing from his understanding of the need for computers to understand meteorology (from his MIT years serving in the military), he would explore computers and technology for what they had to offer the field of education and eventually produce a website in his emeritus years. He writes of this final extension of his "driving force":

> Here was a chance to continue the development of simple investigations by school kids to demonstrate and explain science concepts that I felt were both basic and important, but were not often well understood, and often were not even included in elementary, junior high, or secondary school science. I could continue to create and distribute some of the conceptual developments that I thought kids of upper elementary, junior high, and high school would enjoy. At age 86, I applied for, and was awarded, a Podell grant—my first outright grant ever! (Rockcastle, 2009, p. 233)

Almost 30 investigations were designed for "Rocky's Science Fun." He had a cadre of elementary students who discovered his website with whom he corresponded by email.

Rocky was interested in making the abstract concrete. He engaged the learner with natural phenomena. He modeled the inquiry process. He had Rachel Carson's sense of wonder about the natural world. He had an engineer's curiosity about "How does this work?" And his sense of "looking below the surface" led his explanations of natural phenomena to reach to physical and chemical explanations. He was an intellectual surgeon probing deeply into a phenomenon and an explanation. His strength was in having others experience the student's inquiring role.

CONCLUSION

Verne Rockcastle is a remarkable person. A unique set of personal qualities made him an outstanding teacher and mentor.

- *Interested.* The world around us is full of wonders waiting to be discovered and uncovered, endlessly intriguing. And he could see them, was always on the lookout for them, from an early age.
- *Hardworking and Prolific.* Rocky's work ethic was exemplary. He was in the office before anyone else, usually writing, anything from a science leaflet to a letter to the editor.
- *Meticulous.* A job to be done was a job to be done well. This was evident from the care he took with his boats to the scrutiny with which he reviewed any text being prepared for publication.
- *Generous.* He gave of his time, his attention, and his efforts, to make things better for students and teachers, and for those he encountered in his many roles. His generosity often extended to the personal level.
- *Energetic and passionate.* Whether conducting class, leading a field trip, running a race, or just carrying on a conversation, his enthusiasm and passion bubbled over.
- *Optimistic and inspiring.* You felt that he believed in you, that you could make a difference, and you wanted to make a difference.

Verne Rockcastle made a difference, in the field of science education, in classrooms across the country, in the lives of those he encountered in his many roles, and through the contributions of those he inspired.

In response to the question, "What are you most proud of professionally?" Rocky responded, "Taking the classroom out into the environment and studying things in the environment as part of a regular class" (Weissinger & Rockcastle, 2014).

ACKNOWLEDGMENTS

The authors want to acknowledge Rocky's former students and colleagues who enthusiastically contributed to the chapter.

Jeanette A. Bakke, Faculty member, Bethel Seminary, St. Paul, Minnesota
Bonnie Barr-Larkin, Professor Emerita of Education, State University of New York at Cortland
Ted Bredderman, Associate Professor Emeritus, State University of New York at Albany
Jim Chandler, Consultant Teacher in Science for the Auburn Public School Department, Director of the Auburn Land Lab
Michael Cohen, Professor Emeritus, Indiana University Purdue University Indianapolis
Richard McLeod, Professor Emeritus, Michigan State University
Carol Hardy, Biology Senior Lecturer, Cornell University (retired)
Charles (Chuck) Roth, Director of Education and Interpretation, Massachusetts Audubon Society (retired)

NOTE

1. http://assessment.aaas.org/pages/references.

REFERENCES

Bakke, J., & Rockcastle, V. N. (1967). *Electric charges and circuits. Cornell Science Leaflet, 60*(4). Ithaca, NY: Cornell University College of Agriculture.

Bredderman, T. (1974). *Elementary school science experience and the ability to combine and control variables. Science Education, 58*(4), 457–469.

Driver, R., Squires, A., Rushworth, P., & Wood-Robinson, V. (1994). *Making sense of secondary science: Research into children's ideas.* New York, NY: Routledge.

Gordon, E., & Rockcastle, V. N. (1953). *Air laboratories. Cornell Rural Science Leaflet, 47*(2). Ithaca, NY: Cornell University College of Agriculture.

Horwitz, E. (1968). Sleight of light. *Cornell Science Leaflet, 62*(2). Ithaca, NY: Cornell University College of Agriculture.

Louv, R. (2008). *Last child in the woods: Saving our children from nature deficit disorder.* New York, NY: Workman.

Palmer, E. L. (1949). *Fieldbook of natural history.* New York, NY: Whittlesey House.

Ripple, R. E., & Rockcastle, V. N. (Eds.). (1964). *Piaget rediscovered: Selected papers from the conference on cognitive studies and curriculum development. Journal of Research in Science Teaching, 2*(3), 165–267.

Rockcastle, V. N. (1961). Little climates. *Cornell Science Leaflet, 55.* Ithaca, NY: Cornell University College of Agriculture.

Rockcastle, V. N. (1963). *An elementary science source book.* Bismarck: North Dakota Dept. of Public Instruction.

Rockcastle, V. N. (2009). *Acorn to oak.* Internet-First University Press. Retrieved from http://hdl.handle.net/1813/1417

Rockcastle, V. N., & Bakke, J. (1967). Pond life. *Cornell Science Leaflet, 60*(2). Ithaca, NY: Cornell University College of Agriculture.

Rockcastle, V. N., Horwitz, E., & Harkenrider, J. (1968). Things to do with you. *Cornell Science Leaflet, 62*(1). Ithaca, NY: Cornell University College of Agriculture.

Rockcastle, V. N., Salamon, F. R., Schmidt, V. E., & McKnight, B. J. (1970). *Stem—Space time energy matter—Science.* Menlo Park: Addison-Wesley.

Schmidt, V. E., & Rockcastle, V. N. (1968). *Teaching science with everyday things.* New York, NY: McGraw-Hill.

Wiessinger, J. (Interviewer) & Rockcastle, N. V. (Interviewee). (2014). *A conversation with Verne Rockcastle* [video recording]. Internet-First University Press. Retrieved from http://hdl.handle.net/1813/36390

CHAPTER 11

DR. PAUL KUERBIS

A Pioneer of Small College Science Teacher Education

Peter Veronesi
SUNY–Brockport

INTRODUCTION

When you first meet Paul Kuerbis, you instantly understand why he was asked to be on so many national committees, was nominated for and elected to the presidency of the Association for the Education of Teachers in Science (AETS, which eventually became the Association for Science Teacher Education, ASTE), was so successful at securing such a respectable amount of funding from the National Science Foundation (NSF), and certainly why he fell in love with teaching science in ways that fostered curiosity within his students.

Paul greets you with an affable smile, a quick and gentle wit, and a chuckle that would disarm the most stern among us. Paul is kind, thoughtful, and he has always been a big-picture person, able to see the most critical pieces of the puzzle in issues surrounding science education. From his humble beginnings as a boy on his "Huck Finn raft" floating about and looking for critters in the "wash" near his home in California to the Director of the Crown Center for

Going Back for Our Future II, pages 191–226

Teaching and Learning at Colorado College, Paul's involvement in science and science education has spanned 50+ years and continues to this day.

This chapter is a result of a series of interviews with Paul and is dedicated to him and the many ways in which he contributed to the advancement of science education at the local, state, and national levels. In this work, we will discuss a time span of more than five decades and encompass experiences from his childhood through capstone projects; from significant opportunities to life-changing relationships; from working with Jim Ellis at the Biological Sciences Curriculum Study (BSCS) in the 1980s to their winning a National Association for Research in Science Teaching (NARST) award for practical implications of using microcomputers for professional development (1989) to an ever-increasing leadership involvement into the 1990s with AETS.

Paul's leadership roles started early and continued, for example, by his first becoming president of the Colorado Association of Science Teachers, then the director of the Southwest region for AETS. This kind of professional involvement led Paul to a greater focus on national movements that led him to becoming a strong advocate of professional development for teachers in science education, which involved other key leaders in science education like Rodger Bybee, Susan Loucks-Horsley, and Angelo Collins; through leadership with the Colorado Association of Science Teachers. Throughout all of these experiences, Paul kept riding a wave of professionalism that got noticed when somebody nudged him and said, "Hey Paul, I want to nominate you for the AETS presidency," which he was successful at as well. And the rest is of course, "history."

Those who recall the musical *Dr. Dolittle* will understand the meaning of a Push-Me-Pull-You, truly an animal that seemed to have the strange ability of traveling simultaneously in two directions but has to decide on a single choice to actually get anywhere. Paul Kuerbis, and we suspect many other science education pioneers, made the choices throughout his journey where he saw the best opportunities for success or fruitful outcomes from his endeavors. Indeed, human beings are characterized by being drawn toward things in which they find pleasure or success and move away from those aspects of life we do not. So, where did Paul *not* spend his days? Clearly, one example is the displeasure of grinding up rat uteri. Too much time spent in this awful experience told Paul he was not going to be a bench scientist in a lab. Was he going to bang his head against the wall with the thought of grinding up those mammary organs hour after hour after boring hour? Hardly! So, becoming a bench scientist (as defined in the 1960s) was certainly not to his liking. As an additional and personal assault to his realized boredom, Paul was accused of "not being a good scientist because he didn't like the process of the repetitive nature that was inherent to becoming a bench scientist" of those days. With this example, it is easy to see what were the closely held beliefs of those involved in actually doing science during his early years. Paul was simply not willing to "pay the graduate student dues" to enter the club. So, this aversion to such a repetitive process in doing science was perhaps a huge help in terms of supporting his eventual career satisfaction in teaching science.

As in any life achievements, times and dates are important, but they never follow perfect paths. Nor do they follow decades nicely to make a direct linear sequence, Instead, life follows complex twists and turns and many things have an overlapping function about them. The reader may find it helpful to refer to the Gantt Chart (see Table 11.1) that shows Paul Kuerbis' many pathways and how they overlapped during the decades he spent in science education.

Influencing Childhood Experiences . . . Perhaps

Childhood should be a time of learning and playing, or perhaps, "messing about" with things. In Paul's case, he had experiences, influences, and the harsh reality of chronic and acute asthma and bronchitis, all of which helped to shape his early thinking. Paul was fortunate to grow up on an acre of land in California, where he helped to build the family home with his parents. All kinds of practical notions of science began to become part of his thinking. Pounding nails, sawing boards, and making things level all contributed to the *Engage* and *Explore* phases of Paul's childhood, and as with hands-on science, these multisensory experiences stuck with him throughout his life.

When it was time for real play, one of the greatest adventures Paul had as a child was, after reading about Huck Finn, building his very own Huck Finn raft and setting about exploring the "great wash" next to the family's land. This "great wash" later turned out to be a runoff area for Altadena River in California. In his "messing about" on the raft in the wash, Paul wondered about and explored the "stuff in the water," which included frogs, micro bugs, and the many species of water-loving plants and algae. Back on dry land, Paul continued to learn about biological life by growing vegetables with his dad, finding out the nuances that made things grow. Similar to what happens to many children, Paul's confidence and wonder grew in ways that led him into trapping gophers, which eventually led to gopher dissection and a keen interest in becoming a medical doctor.

Since Paul had very serious health issues as a child, he spent a lot of time in the doctor's office and he began "knowing" he was going to become a doctor. It was these critically important doctor visits to deal with his asthma and chronic bronchitis that reinforced his continued drive to learn more about science. These experiences are a snapshot of what we might call Paul's informal science education that helped to guide him toward a continued interest in wondering about all science-related things.

While Paul's informal science education was unstructured yet meaningful, his formal science instruction was, for the most part, less than desirable. In an example that has become far too common, Paul was offered very little science in his K–8 education. In fact, one particularly bad experience in sixth grade had Paul "reading about science" in a very hot southern California classroom, causing him to faint and taking the desk with him as gravity worked its magic, pulling them both to the floor.

Most memories about high school science were similarly dim. One of Paul's science experiences was a biology course focusing on the memorization of the names and definitions of organisms. Chemistry, not faring much better, was taught by an older man whose background came from the chemical industry and who religiously followed the text, which in the 1950s focused on manufacturing. As many stories are used as analogies in a chemistry class, the stories told in Paul's chemistry class were replete with examples about chemistry manufacturing details like the making of cement, which made sense since making things was in the life experience of the chemistry teacher.

Physics, on the other hand, was experimental, probably the first of the reform approaches that Paul can recall. Doing investigations was what Paul experienced in his physics class. "Doing science" contrasted greatly with his chemistry and biology coursework, which was almost totally note driven and lecture based. In fact, Paul did not know his physics teacher was avant garde at the time in terms of hands-on, minds-on science.

CRITICAL INFLUENCES

Paul Kuerbis describes the people who were the most critical influences in his life as individuals who unknowingly, yet absolutely, guided him toward his eventual career in science education. These include in order:

1. Paul's Mother:"You WILL go to college!" making Paul a first-generation college student.
2. The "New Guy": An avant-garde, new biology faculty member at UCLA-circa 1965
3. A buddy: A senior TA who invited Paul to try teaching for the very first time. "I fell in love with teaching!"
4. "No jobs in sight!": so... David Bartine, a fellow high school teacher who talked Paul into a possible "backup plan" to get a PhD in science education at Berkeley.

With each of these individuals, Paul furthered his career path toward science and his own personal journey toward the teaching of science.

College #1: The Undergraduate Experience

When Paul left high school and entered college, he was told very quickly by one of his organic chemistry professors an interesting tidbit of information. His college chemistry professor said, "I want you all to know that everything I had in my master's degree in 1938 you will have, and more, in this single year-long chemistry course" (P. Kuerbis, personal communication, August 7, 2014).

It isn't clear to Paul why such a statement was made. Was this intimidation? Was this a weed-out course? No one will ever know, but it gave Paul a healthy fear, which focused his energies on studying. This professor had a collection of jars with leftover synthesis products, and Paul and his peers were allowed to "play with chemicals" all the time and gained confidence that they could do no harm. Benzene? NO PROBLEM! They even bragged about the holes in their lab coats that resulted from the acids they worked with freely. In spite of the "wow," biology continued to be a "love of learning." His recollections about some of the facts of the day included hearing that "we did not know if humans had 46 or 48 chromosomes," as the textbook they were issued was one on classic 1959 taxonomic biology.

One of Paul's early influences in college was Brother Lawrence Cory, who was very excited about the "emerging information" on large molecules like DNA, proteins, and enzymes, which was apparently not an easy course for the 30 young men in the class. Paul subsequently earned a "D" at midterm, but increased that to one of only five "Bs," the highest grades in the class, by the end of the course. Embryology was another course in Paul's education that he found fascinating. In this course, they learned about early cell layers, in which Paul still vividly recalled a picture of a frog leg coming out of a frog's back that was produced from stimulating undifferentiated cells. Comparative anatomy, on the other hand, was frustrating as he and his friend noted early discrepancies and contradictions between information in the text and the information from the professor. In one instance of pointing out these contradictions during class, the professor proceeded to "put Paul down" publically for asking questions in class. This public humiliation resulted unsurprisingly in Paul asking no more questions. As a result, Paul and his friend ended up typing their class notes along with notes from the text to get the best and most accurate information, earning both Paul and his classmate an "A."

Brother Lawrence, the biology teacher mentioned earlier, also taught the evolution course, which was enlightening and which meshed well with a history of science course Paul took while in graduate school at UCLA. Clearly, evolution was a critically important part of biology and helped Paul see how things made sense within the living world. With all of these courses behind him, a "new guy on the block" in biology became a trusted adviser to Paul and recommended he attend his alma mater (UCLA) to earn a Master of Arts in biology.

Onward and Upward: Toward the MA at UCLA, but Grinding Up Rat Uteri?

With an undergraduate degree in biology in hand, Paul was awarded a research assistantship, where he learned in short order that he was in no way cut out to be a dedicated bench biologist. Bored to death, repeatedly grinding up rat uteri for hormone experiments was not in the cards for Paul Kuerbis. Then, at a critical juncture, a trusted friend and more senior graduate student convinced Paul to join him as a TA in year two for the basic anatomy and physiology course. It was during his teaching as a TA at UCLA then, that Paul fell in love with teaching science! As he states,"I was having so much fun helping freshmen learn through hands-on experiments!" (P. Kuerbis, personal communication, August 7, 2014).

In that second year, Paul went on to pass his qualifying exams at the MA level and then started dabbling in science education. He was hooked! Having never taught anything in his life, his experiences as a TA were significant enough to act as both the catalyst and the turning point in his life toward teaching. Paul soon stopped thinking about becoming a medical doctor and began his focus on teaching science instead.

As he was preparing to leave UCLA, Paul was able to sign up for a science methods class that was taught by a high school science teacher who happened to be a proponent of "science as inquiry." From this course, Paul learned a great deal about teaching using science inquiry, and it further pulled him into science education as a career path. With what was then a sound understanding of teaching and learning, and with a strong biological content background, Paul aimed at teaching at community colleges in southern California. However, his lack of experience and there being only three jobs in the area caused Paul to set his sights on a broader array of opportunities.

THE LIL' ENGINE THAT COULD: GETTING ON THE SCIENCE EDUCATION TRAIN AND NEVER GETTING OFF!

Persistent Persistence!

Paul Kuerbis seemed to have what one could say was a continued, unwavering drive to start in science and then stay in science education. Being around and working closely with like-minded people was a key determiner of involvement, growth, success stories, and eventual leadership for Paul.

For example, during one spring break while he was earning his master's degree, he began looking for a part-time teaching position. With a baby due in May, and no jobs in sight at community colleges, he noticed a job posting as a science teacher for that last part of a spring quarter at an independent school—the Oakwood School. He was interviewed and hired immediately to teach 8th- and 10th-grade biology for the last 3 months of the semester. This teaching opportunity opened the door to what would become Paul's daily experiences with teaching science to curious minds.

> They liked me and I liked the kids! There were small classes of 16–20 in portables. In this school, teachers had wonderful freedoms and responsibilities. I think they trusted me to teach so when fall came around I accepted a full-time teaching offer in that school for 1966–67. I had a load of mostly biology but also an 8th-grade earth science and a 7th-grade physical science (boys only). I drove 30 minutes to school, against traffic, while planning and debriefing the day coming back home into town, once again against traffic. I did not realize until many years later how practicing the questions I wanted to ask students the next day and debriefing the day's teaching, just thinking about my day on the way home was so significant and thinking about how many moves I had to change, mistakes I made in responding to kids and the like, was so important to my development as a teacher. (P. Kuerbis, personal communication, August 7, 2014)

As a very new science teacher, this was Paul's recollection of his feelings and sentiment in the mid-1960s. He knew he was hired as a replacement for a teacher who had left the classroom unexpectedly before the end of the academic year in 1966 and he had to prove himself as an effective teacher. How he would do that would change quite significantly and very early in his teaching career.

Driving Against Traffic and Unknowingly Practicing Wait Time

> When I say [I fell in love with teaching], it sounds kind of scary! Getting on a great train that never stops for you. Actually though, I'm not sure of any way else to explain it at the moment. If you're riding on a wonderful train, with excellent sleeping accommodations, exquisite food, and the scenery passing by was incredible, it would just seem to be way too difficult to get off. (P. Kuerbis, personal communication, August 7, 2014)

In Paul's new life as a first-year biology teacher, his daily drive to and from school gave him the opportunity to reflect on student responses that day and how these responses and his actions changed the nature of interactions in his classes. Unwittingly, Paul was practicing what we now know as being a *reflective practitioner*, the development of learning about teaching via constructivism, developing higher-level questions, and even using wait time, all elements of a highly effective teacher that would decades later be shown to be best practices through science education research.

As we now know, we must "know our students" and their prior knowledge and base our instruction accordingly. In a subconscious-conscious way, Paul's intuitive reflection on his questions would lead to much more meaningful and robust student-student and student-teacher interactions in his classroom each successive day. So, from behind the wheel of his southern California 1963 Le Mans convertible, Paul Kuerbis was practicing researched-based teaching practices long before they became commonplace in research literature. And these practices produced the desired success he envisioned for his biology classroom and kept him on the train and on his journey through science education. "Can you believe they sent me to NSTA? I was only a 1st-year teacher! Meeting Jack Carter and Harry Wong at NSTA in 1966 was powerful stuff! They sent this brand new teacher to NSTA! Unbelievable! Yet they did" (P. Kuerbis, personal communication, August 7, 2014).

Paul's early science teaching was like a rocketship ride of adventure and responsibility. At NSTA he was tasked with choosing the most current science textbooks as the school was adding a new grade level each year. He made the decision to switch from BSCS Yellow to Blue version, adopted other cutting edge materials of the day like Earth Science Curriculum Project (ESCP), Introductory Physical Science (IPS), and then the revised ChemStudy and Harvard Project Physics all while he was mentoring a new assistant professor from UCLA who drove out to Oakwood to teach physics in the school. Paul was also responsible for leading the planning and then opening of a new science building at his school the following year. So, by default he was cast into leadership positions very early in his teaching career.

> I would say, "Okay kids! 30 chapters to cover this year! Let's get started!" Egads! Was I damaging kids that first year? I had to switch what I was doing. It was terrible, and I knew it! So, this is a time where I guess I learned less is more. (P. Kuerbis, personal communication, August 7, 2014)

Sometimes known as the "alphabet soup" years of science curriculum, Paul was influenced by BSCS Blue and Yellow versions. Exposure to ESCP, IPS, ChemStudy, and Harvard Project Physics supported Paul's continuing enjoyment for teaching science. Similarly, known today as the guided inquiry method of teaching science became Paul's way of teaching and he loved it! So the transition was complete: from a teacher who thought covering the chapters was the best thing for his students to one who became convinced that *exploration of science phenomena was the way to teach science.* In fact, one of Paul's most memorable lessons was having his 10th-grade biology class looking through microscopes and observing fertilized frog eggs divide. The result of these observations included "tons" of other incredible student engagement and questions as follow-up.

Guided inquiry was a new approach at the time and it matched Paul's progressive vision for teaching science. Paul and his students had a wonderful time exploring reproductive biology by observing the cleavage planes divide in fertilized frog eggs and other associated phenomena. *Scientific American* articles by Lederberg on how genes work and the transcription of DNA by ribosomes were used in Paul's 10th-grade biology class, furthering the content knowledge for Paul and his students. It was a very exciting time! Fruit fly genetics required Paul to make his own recipe for fruit fly medium in his humble apartment, where he raised new broods of flies for his students to use in their explorations. He also created an advanced senior-level biology course that was all labs, discussion, and experiments as they used the BSCS "black" version that had BSCS lab blocks. Indeed, with music playing at a low volume from the radio all the while, lots of labs meant more time for follow-up "what happened?" discussions and less time devoted to lecturing. After Paul began to change his teaching practice from lecture to having his students doing labs, he and his students were able to unpack lab data while also giving up the previous year's traditional schedule of "getting through the book," which never happened again.

Teaching experiences by doing labs and experiments such as these solidified Paul's view of inquiry in teaching science-through-doing that later carried him through his entire career. This "newly discovered" way of teaching science was further supported by his students doing superbly on the national test they were required to take. The scores of the students were high despite the fact that they only covered two thirds of the BSCS Blue version textbook. After his students showed equal or better gains on the national science test without covering the content as in past years, Paul never went back to "covering science as textbook facts." Instead, his newly discovered way of teaching was now focused on "uncovering" the mysteries of the natural world through the inquires he and his students engaged in.

The Oakwood School years (1966–1969) provided pivotal moments in Paul's professional life. In only his second year, he suddenly found himself as the science department chair with the responsibility of mentoring two new-hires who had MA degrees in science but no teaching experience (one from Harvard and one from Yale). He was also in charge of choosing a science curriculum, designing a new building, all of this with very little professional experience under his belt. In such a free, independent environment of that school, Paul was able to thrive. With such freedom at this independent school, Paul continued to "fall in love with teaching science" and with preparing and supporting new teachers. This supportive feeling and other interactions with one of his "trusted others" led Paul to eventually pursue a PhD program. So, in 1969 Paul Kuerbis sought entry into the PhD program at Berkeley and was accepted.

The politics of the day were huge in California and navigating through this politics was an arduous task. Paul's political contribution at that time was to fight for and urge his Oakwood faculty senate to push for equity in pay for women teachers, who clearly were second-tier citizens in terms of pay at that school. Other issues arose that indicated Paul's need to keep his feet grounded in a more secure environment, so he developed a back-up plan so that he and his family would keep a funding stream intact. Even with no graduate student funding during that time, Paul still decided to go forward with his plans for a PhD program. This move required a great deal of courage and fortitude to accomplish.

With advice and career suggestions from his trusted friend and colleague David Bartine, who taught in Oakwood's English department and who was also an alumni of Berkeley, Paul became convinced to check out Berkeley for the possibilities of his earning a PhD. Paul's research into Berkeley led him into science teacher education. So, during one pivotal March spring break from the Oakwood School, he met with science educator David Miller, who had a slot for a PhD graduate student that fall, and David took Paul in. During this interview, the essence of their conversation about teaching and inquiry was mutual, so another professional bond and watershed moment happened in Paul Kuerbis' life and the PhD was next on the list of goals.

With most of the major 1960s protests on campus behind them, Paul was again able to focus on his initial experiences in learning more about the most effective practices of science education. His first K–6 science methods class was with Professor Jim Oakey, who was a visitor and expert on the processes of science. Apparently Oakey did nothing but science inquiry and model the use of science processes. In fact, he assigned Paul and his contemporaries such questions as, "Does hot water or cold water freeze faster?" Trying to answer questions like this was the core of Paul's first experiences while earning his PhD in science education at Berkeley.

As is it today, earning a PhD was not an easy task to complete for Paul Kuerbis. From 1969 to 1976, Paul was to take a job and struggle to complete the terminal degree. All the while, his family increased to a total of five members who were counting on him to succeed. Paul had the challenge of trying to find the balance of family life and his graduate work. Back at the Oakwood School, Paul and David would continue to be tennis bums and also read and talk about common materials such as Noam Chomsky's writings on linguistics and deep language and Piaget's various writings on language development and logical reasoning.

Cutting and Pasting; No Really, Cutting and Pasting!

During the days at Berkeley, Paul continued to grow and expand his horizons in professionalism and working with others, which turned out to be one of his greatest strengths as a Pioneer of Science Education. Key collaborators and peers at this time were Mary Collins and Rich Ponzio, who were working on K–6 learning centers and introduced Paul to using the Myers-Briggs type indicator with teachers and administrators. With this group and their faculty advisors, David Miller and Larry Lowery, Paul began helping to write, and yes, literally cut and paste together an NSF grant proposal, which was later funded. There is a reason that Microsoft and others call it copy/cut/paste when we word process these days. It is because in the days prior to white-out, one actually did have to use scissors, tape, and glue to bring thoughts together on paper in a way that would make better sense to readers.

Cutting and pasting completed, this first grant fostered in a new approach to teacher education and an early version of K–12 team-taught teacher preparation that was based in public schools. Paul also had an incredible opportunity fall in his lap during his first winter at Berkeley, working at a newly developed institution that was in its infancy—the Lawrence Hall of Science. Not only was it a great way to earn money, Paul was also given the honor and responsibility of establishing the very first Professional Development Resource Library for staff and visitors at the Hall. He had to decide what should go into it. Which journals? Books? All of which was a rushed, 3-week learning experience. Paul was again fortunate to have the energy and creative ideas to help establish the resource library of this new institution that was just beginning to develop outreach programs for teachers beyond the successful K–6 Science Curriculum Improvement Study (SCIS). He was invited to choose the journals and other print resources that staff and teachers would interact with, and other artifacts that would have the greatest influence on their professional development.

Paul was simultaneously involved during 2 years as a part-time "supervisor" of K–12 student teaching, and in his final year, he was hired full-time as a college supervisor in secondary science, which had him making more money than his Berkeley advisor! On the academic side of things, Paul had what he termed "great unofficial and official mentors," like Larry Lowery, who got him involved in researching Piagetian levels in children and who later became the founder of Full Option Science System (FOSS); Robert Karplus, physicist and founder of SCIS; Mac Laetch, head of the Botantical Garden, and his outside content person lined up for his eventual oral defense. In an interesting twist, it was Mac who asked Paul, "What do you think is the major idea holding biology together?" Paul did not realize that just months earlier, Dr. Theodore Dobshansky had published a treatise on evolution as the central idea. With gentle hints from his mentors and recalling his course on evolution and the new genetics he had taught at Oakwood, he came to the same conclusion as Dobshansky. Always challenging him to excel and move forward, working informally and formally with these individuals at Berkeley simply created a plethora of tremendous experiences and opportunities for Paul in his fledgling career in science education.

Struggling nonetheless to carry out his research, collecting interview data, all while pushing forward with dissertation drafts, he applied for the only two known positions in science teacher education. Both positions were at recognized liberal arts colleges, settings that resonated with Paul. Paul interviewed for both over a 5-day period and was offered them both. Now history will show that he turned down one and chose the other at Colorado College (CC). At CC, Paul could teach both elementary and secondary science. CC was where Paul began and remained throughout his entire professional career.

Colorado College, Colorado Springs, Colorado (CC)

Paul arrived at Colorado College in late August 1973 as one of two members of an education department, which was a rather unique situation for someone prepared at a major university where the expectation was to conduct and publish original research. In addition, you were expected to focus on preparing future graduate students as your primary goal. Teacher education was not necessarily in the cards. Paul's personal adventure was common to professionals of the day. His trek from Berkeley to Colorado was fraught with anxiety. A new job to go to, thinking about an unfinished pie of dissertation materials housed in a large box, wondering if his U-Haul truck of furniture, houseplants and other goods would find enough gasoline during the country's first embargo of oil and shortage of gasoline. The Autoclub, a precursor to Triple A, was relied upon for the best information on where he might find gasoline along I-80. Because he was alone, driving the U-Haul for 4 days, he was able to use the quiet driving time to think about teaching and to mentally construct a method of teaching secondary science methods courses that were coming his way in the upcoming Block Plan at CC. It was 4 days that really brought back memories of his days teaching at Oakwood and using his driving time there to reflect, think, and plan in his drive to work.

Teaching and advising took center stage for Paul at CC in those early years, and work on his dissertation fell all too easily into the shadows. And and he gave little if any thought to attending meetings of professional organizations. At that time, Paul was only vaguely aware of NARST and knew absolutely nothing of AETS. So there was really no reason for him to become associated with these organizations until they came into a greater light for him, which was not to happen until a decade later. In fact one of his advisors remarked that NARST was oriented more to the eastern part of the country and very few from the West Coast seemed to be involved. In fact Paul never had discussions about why, as a new faculty, he should even consider involvement in any professional organization. And furthermore, most articles in *JRST* were from the Midwest or East, certainly the West did not seem represented. After all this lack of interest in becoming involved in any professional organization, it's not exactly clear how Paul found out about a January 1974 meeting of the Southwest region of AETS, but he did and off he went to the conference in his tiny VW Squareback where he made it to the YMCA lodge in Estes Park in early January with snow beginning to fall all around.

Unfortunately, Paul had lost track of SW-AETS in subsequent years and was not aware of any STEM colleagues in Colorado who were involved or participated. And it was much too far of a drive to meetings in Kansas, Oklahoma, or Texas, especially in January. Eventually though, Paul did manage to attend a few NARST meetings and, while he found them intellectually interesting, the culture at that time seemed to be similar to that of a "one-upmanship" during concurrent sessions. This was contrasted and very different from the nurturing fireside chats and concurrent sessions with graduate students and faculty coequals he found at AETS meetings.

Moving on from his doctoral work at Berkeley, even though not completed, CC in Colorado Springs, Colorado, offered Paul Kuerbis a brand new professional opportunity. CC, at the base of Pike's Peak, is an institution that gave Paul his next challenges, goals, and further access to higher education that led to 40+ years of opportunities that he both created and acted upon as they came his way. Having CC as his base provided a relatively secure professional development platform for science teacher education, national involvement through AETS, NSF grants, and continuous professional development for science teachers in and around Colorado until and through his retirement in 2012.

Paul once again began making significant curricular decisions immediately at CC, where the block plan of teaching one course at a time allowed him to place students in schools each morning and have class on campus in the afternoon. This was a time when many across the nation were criticizing teacher programs for not giving candidates enough experiences early in their preparation, so Paul found himself poised well ahead of the rest of the profession. It was a sign of how Paul would continue to see the bigger picture and act upon that vision.

At the same time, he began to focus on writing his first successful "very small" NSF proposal to work with area high school science teachers and CC science faculty to strengthen their science backgrounds, especially in field biology and geology, and a new geometry called topology. Paul's grant was able to serve 200 teachers for less than $20K. Writing the proposal gave him a taste of how to construct convincing arguments for ideas he believed were important.

Paul sought additional funding in 1980–1981 to purchase two Apple IIe's, complete with 9-inch green screen monitors and the most advanced printer of the day, the "dot matrix printer," all with the goal of getting teacher candidates used to technology. In many ways, this initial campus-based grant was novel in that there simply was no technology available for teacher candidates at the time. There was also absolutely nothing in terms of technology in any Colorado schools at the time, so it was clear why this CC-based grant was funded. Being ahead of the curve with the Apple IIs reinforced Paul's growing savvy in making strong cases for accessing funding.

Revisit the fact that Paul left Berkeley with an unfinished doctorate. Doing so led to a real threat by his then–department chair and dean to complete his dissertation by the end of his third year, 1976. In Paul's case, one of his advisors, physicist Robert Karplus (Bob), suddenly demanded a sixth chapter, which were the actual results of his study of high school science teachers. Bob wanted a qualitative interpretation of their attitudes. This additional work put an incredible scholarly burden on Paul by making his dissertation more than a straightforward, simple instrument design of attitudes toward instruction using projective techniques. With this additional requirement to Paul's dissertation, Bob agreed to read revised drafts as he received them from Paul via courier. Since the Internet had not been invented yet, Emory Air, a fledgling company predating Federal Express, was chosen as the most direct vehicle for written communication. So Paul set about sending drafts of his revised dissertation overnight via Emory. Once each successive draft had been received and read by Bob, the two arranged times that they could talk on the phone and discuss revisions to the dissertation. Paul would take notes as Bob dictated them over the phone to him, which usually happened while Bob was eating his breakfast.

To further help the process of writing along, Paul's typist allowed for future pages since electronic word processors, like the Internet, had not yet been invented. Fortunately, Paul was indeed able to complete and file his dissertation with little time to spare. And in doing so, it freed his professional time for further explorations in the national involvements that would become his future.

His dissertation behind him, with his new-found freedom, Paul was able to take on the new responsibility of teaching the elementary math methods at CC in the late 1970s while simultaneously becoming involved in preparing to support local schools who were adopting developing mathematical processes. At this time, Rand McNally Publishing supported a week at Madison, Wisconsin, with the program developers. With this new training, Paul was able to go deeper into learning-centered and research-based instruction. He instantly saw the wisdom of the program's nontraditional approach of using small groups, no text, and teaching for understanding. But it was also a time at the national level of resistance to new math and a push to return to "the basics" with memorization. To learn more about the Rand McNally–type programs, Paul, in what continues to be his characteristic way, began attending National Council of Teachers of Mathematics (NCTM) meetings to strengthen his math teaching background. In his words, "Keeping the faith of inquiry learning, Socratic dialogue, and what might now be described as constructivist philosophy" were notions that drew him into teaching in ways he had begun back in California. And while this new learning continued to prepare him almost solely to engage in preservice work, he also recognized his need to develop expertise in lifelong cross-curricular teacher development as well.

> Colorado College gave me all the opportunity in the world to teach science content courses and be involved in elementary and secondary science teacher education, they added on mathematics education work and I challenged myself to push the envelope of more effective work with experienced teachers, so I stayed. (P. Kuerbis, personal communication, August 7, 2014)

The 1980s: A Steady Stream of Work at a Small College

"Reagan cuts all NSF funding. People got their grants pulled and there was simply no money for grants in Science Education" (P. Kuerbis, personal communication, August 7, 2014).

And so started the decade of little or no funding for science educators in the United States. It was certainly a time when everyone had to dig in and carry on against a wall of adversity. However, as with every serious dip in funding, there has always been a rise on the other side. And so a light emerged toward the middle and end of the 1980s for everyone interested in pursuing grant funding, and Paul Kuerbis was among that group. This was a proposal-rich time, where Paul began to work with Jim Ellis, the new science educator at BSCS and a fellow who was to become a large part of Paul's professional life at Colorado College. Paul had recognized the importance of collaborating with colleagues at the college and now he saw the opportunity to work with Jim, whose logical sequential style was a perfect complement to Paul's more abstract random and big picture style. They would come to know how they needed each other. They learned how such a team, each member respecting the strengths and weaknesses of the other, could be leveraged. Over time, Jim would precede Paul in becoming both the director of Southwest-AETS and then president of AETS.

> We were a great team, Jim Ellis and I! Teaming up with Jim created an incredible synergy. We started with a NSF-funded study of what teachers needed in order to be computer literate. That led to two more grants: one developing materials for computer literacy and then one studying the implementation of those materials. (P. Kuerbis, personal communication, August 7, 2014)

Paul was the point person on teacher development, with Jim focusing more on the organizational side and budgets. The new team of Jim Ellis and Paul Kuerbis collaborated on all aspects of grants. Then in the early 1990s, they took what they had learned from all of their work on ENLIST Micros and collaborated on the NSF-funded Colorado-Science Teacher Enhancement Project (CO-STEP). Six sites, including one at CC, were chosen to engage in a program that studied how teachers could deepen both their science knowledge and their pedagogy; it was a lot of work and a great deal of fun.

Then suddenly in 1983, a groundbreaking and "new paradigm-starting" publication was released entitled, *A Nation at Risk: The Imperative for Educational Reform.* Paul would later argue that it was this seminal national publication event that led him to his eventual leadership roles in science education and his fundamental involvement at the national level over the next 30 years. As a result of *Nation at Risk, Science for All Americans* (*SFAA*) was subsequently published by the American Association for the Advancement of Science (AAAS, 1989). It was then that *SFAA* formed the platform for *The Benchmarks for Scientific Literacy* (AAAS, 1993). All of the hullabaloo from *A Nation at Risk* forced the U.S. Senate to approve funding of the National Center for the Improvement of Science Education (NCISE). And so, moving with the incoming tide, Paul's work became increasingly geared toward educating science teachers (preservice and inservice) for the new millennium. That tide would eventually wane, but not recede for many years. For Paul, it culminated in 2000 with a $6 million, five-year NSF grant called Science Teacher Enhancement Project Unifying the Pikes Peak region (STEP-Up). This work demonstrated that effective professional development, along with teachers truly adopting active learning pedagogies, could impact student achievement in reading, writing, mathematics, and science.

1988: A Flood of New Responsibilities

With well more than a decade of growing abilities, confidence, and vision within him in and continued involvement in advancing science education, Paul began moving from his beginnings as a presenter at science education conferences like SW-AETS and NSTA to writing grants and to ever-increasing involvement with AETS. Paul's increasingly motivated involvement led him to what might be called another pinnacle when he began continuous work as a natural on national advisory boards in the late 1980s and throughout the 1990s.

> One of the grants I wrote provided the opportunity for K–12 teachers of science to talk to each other. Each shared the struggles of teaching within their own populations. It was a very positive time getting these populations of teachers to share insights and concerns with each other about teaching science to their own student cohorts. I should note, that everything we did during class was from a researched-based pedagogy. It was incredibly challenging to keep 3 to 4 cohorts working simultaneously. The 1st 15 to 20 was manageable, but this ended up in a group of 40 one summer, which was far too many to manage effectively. (P. Kuerbis, personal communication, August 7, 2014)

To endure NSF funding issues, the Biological Science Curriculum Study (BSCS) was downsizing, sold its building near Boulder, and reached out to Jack Carter (their Associate Director in the 1960s) to step in as half-time executive director. When the College approved, they asked Jack to see if CC would provide space on its campus, which they did. It was conveniently located out the back door of the College's Education Department in Mierow House. Thus, it was very easy for Paul to begin interacting with Jack and the small staff of BSCS. Paul participated in helping them regroup, eventually interviewing two subsequent hires, Nancy Landes and then Jim Ellis. After 3 years, Jack (once again) needed to return to full-time teaching in the biology department. BSCS, noting NSF funds were once again available and growth possible, hired one of their own to lead them, Joe McInerney. Joe quickly recruited Rodger Bybee from Carleton College to become the full-time Associate Director. By then, ENLIST Micros had been funded initially and Rodger encouraged Jim and Paul to look forward to proposals that would expand that work.

Paul had known Rodger through meetings of education faculty within the 12 small colleges of the Associated Colleges of the Midwest. Carleton and Colorado College were both members. Paul had attended some of Rodger's sessions at NARST. It was Rodger who continued to cultivate Paul's involvement with BSCS and also solidify the presence of BSCS on the CC campus. It was through BSCS that Paul, serving on one of their advisory boards, met Susan Loucks-Horsley during the summer, before NCISE was funded for a partnership between the Network (Andover, MA) and BSCS. It was December of 1988. For that same academic year, Rodger had convinced Paul to apply for and take a full-year sabbatical and become involved in helping BSCS plan the professional development portion of a new K-6 innovative curriculum supported by an NSF grant.

As a result of being awarded a sabbatical in January 1989, Paul shifted his work to the NCISE project. NCISE had three divisions: curriculum/instruction, assessment, and professional development. Rodger chaired a working group on curriculum based in Colorado Springs, Senta Raizen chaired the assessment group in Washington, DC, and Susan led the professional development group in Andover. Paul was asked to serve a role similar to a project director (without title) and provide the "glue" across the three divisions by participating in all meetings of the three groups. The plan was to focus in year one on K–6 science and produce three monographs, one for each division. In the second year, the middle-level science would be the focus with the production of three more monographs. Then, in year three, the secondary level. Paul would co-lead the planned publications with the three chairs. Included in this group of science educators were Audrey Champagne, Ed Haertel, Ina Mullis, and Senta Raizen in the assessment group; Sally Crissman, David Heil, and Rodger Bybee for curriculum/instruction; and Susan Loucks-Horsley, Jackie Brooks, and Karen Worth on professional development. From Paul's perspective, the people involved in all the committees were stellar individuals who he had only read about or who he barely knew.

Paul learned a great deal and also contributed to the many insights contained in all of the NCISE monographs. In one instance, Paul proposed to Susan, Jackie, and Karen the idea that the constructivism learning notions being discussed in the curriculum/instruction group could provide a framework for teachers' professional development. "Shouldn't teachers, like their students, be actively constructing knowledge and skills about their teaching?" he asked the group. It was in fact an approach Paul had found himself naturally taking with his work in science and math teaching courses and student teaching supervision at CC. Paul fondly recalled writing so many then-novel ideas on behalf of the curriculum group, which Rodger Bybee chaired. Rodger's expertise was particularly in science content and conceptual frameworks. The 5E instructional approach he then was toying with triggered another important event.

Paul and David Heil conducted an interview with one of their panel members, geologist Ed Buchwald from Carleton College. Ed recounted how he taught like he conducted his science investigations. In his first day of class, for example, he took students to a field area on campus and had them record their observations. It was his way of inviting them into earth science and allowing them to discover and explore what they noticed. The result of the interview was the curriculum and instruction panel creating its own instructional model referred to as the Teaching Model in their monographs, including the final singular monograph on high school science.

Paul concentrated on those sections dealing with teaching and learning. In addition to the Teaching Model, he was successful in crafting a new definition of curriculum in chapter four of *The High Stakes of High School Science*, the single monograph the project committees produced in 1991 (NCISE, 1991).

> Curriculum is a dynamic enterprise, perhaps best defined as a program of activities undertaken by students and teachers, alike. To reflect the Center's new view of instruction, perhaps a new term is needed: curriculum and co-instruction, or more simply, construction. The term construction implied that teachers and students engaged in the kind of activity together that ultimately resulted in a generation of new understanding. At the very least, the Center insisted on viewing curriculum and instruction as different sides of the same coin. (pp. 79–80)

Paul was disappointed that this proposed definition never gained traction, caught on, or even caused a mild ruckus. He had to be content that the NCISE groups agreed and that maybe someday his big-picture ideas would catch on.

A tribute to the cutting-edge work of NCISE, however, was that many of the NCISE participants went on to take prominent roles in guiding the three working groups for the National Science Education Standards: Rodger, who chaired the curriculum working group; Karen, who chaired the Teaching Working Group (on which Paul served); and Audrey, who chaired Assessment. And some members of the NCISE advisory board also were tapped for work on NSES. It later became clear that the NCISE monographs became the basis for much of what subsequently appeared in the *National Science Education Standards* publication of 1996.

It was clear to Paul in early spring of 1989 as NCISE was just starting, he would soon be asking his department chair and dean of CC to continue his leave. They granted him half-time leave to complete the 3 years of NCISE responsibilities, which made it clear that without the support of time off regular duties, advancement of ideas would be difficult to achieve. Then, as Paul ended his NCISE work, he returned to full-time teaching responsibilities in his small department at CC.

At that time, Paul was one of only two full-time faculty, so he knew full well how difficult it is for science educators whose responsibilities it was to teach, let alone thrive, in such a small institution of higher learning. According to his chair, the department had "limped along" in his absence, with a visitor from a nearby university filling in for his science education responsibilities and new part-timer taking on the mathematics education work. In no uncertain terms, Paul was informed that he was needed to continue to direct the secondary program and continue what he had never given up: summer responsibilities teaching Master of Arts in Teaching Elementary School prospective teachers and running a highly successful summer science program for 200+ youngsters. It was that "lab" he had created that was critically important in ensuring the MAT candidates learned how to use inquiry as a teaching approach under the guidance of mentor teachers. That intensive 4-week (2 weeks for the children) program is still in place at this writing, a testimony to Paul's big-picture thinking and foresight. Of course, the program also followed Paul's IDEA 2 model—Invite, Discover, Elaborate, Action with continuous Assessment model—an initial week of investigations and discovery activities for the students, a week of making sense of it all via actual teaching and implementation of the inquiries, then a final week of *reflection* and production of units of study to implement in their student teaching and internship classrooms.

In addition to the experiences already mentioned, the decade of the 1980s was a fruitful time in other ways for Paul. This was the time when Paul started his involvement with the Colorado Association Science Teachers (CAST), where he eventually became the president as well as the chair of their convention, reviving the defunct annual conference. Although initially delayed by a massive March snowstorm, causing Paul to cancel the conference, he led his team to regroup and carry off a financially successful May meeting. It produced a substantial net profit to ensure CAST would continue. This example in the mid-1980s is one of the first to show how Paul Kuerbis' leadership potential was reaching fruition and giving him the confidence to step up and lead. His involvement in CAST is one way of demonstrating how his persistence for working toward goals helped other professionals grow in their craft. CAST continues to thrive today. As history would show after Paul's involvement with CAST, he went on to co-chair the program for a regional NSTA conference in Denver in the late 1980s.

As Paul continued his work at the state level, those around him noticed and pushed him toward involvement at the national level. Following is a comment that was shared by a colleague that summarizes a rationale for Paul being selected for involvement in opportunities such NCISE, membership on national advisory boards, and nomination and later presidency of AETS: "I think Paul is characterized by starting progressive science education learning centers and ensuring their successful outcomes. In fact, quiet but solid leadership is in his soul. Paul Kuerbis is the glue that brings and holds committees together!" (P. Kuerbis, personal communication, August 7, 2014).

This kind of personal sentiment from a colleague is also reflected in the opening introduction of this chapter and certainly seems to be a central theme in Paul's professional life. In fact, Paul continued to carry on in a positive direction by pursuing further grants at NSF. It was soon after receiving an initial NSF ENLIST grant that Paul and Jim Ellis began to get on the speakers' circuit at national and regional NSTA meetings. It is also the time, in the mid to late 1980s, that Paul's major work with AETS and his growing national involvement began. In fact, Pauls' earliest memories of SW-AETS "interactions" were once again of a personal/professional nature. Paul's very first AETS meeting was ushered in by another massive snowstorm in Estes Park, Colorado. His roommates at that time included Bobby Jones and other AETS elders from the SW-AETS region. Paul would remark that Bobby's humor and insights into teacher education might never again be found in a single character. It was initial personal connections like these that provided the welcome opportunity to "sit around evening fires and enjoying some delicious red wine and other beverages."

It was Jim Ellis, then at BSCS and constantly on the CC campus and who had been involved in SW-AETS as a graduate student at the University of Kansas, who nudged Paul to join him in participating again in SW-AETS. These regional meetings gave them a venue to share their work on ENLIST Micros and provided a terrific opportunity for senior faculty to introduce their graduate students to sharing research. In fact, many would travel in vans and share rooms so they could attend the meetings at low cost. The gatherings were simply a wonderful time to make connections and network. In the late 1980s, Jim served as president of the regional group, then Paul also stepped in to serve in that role and lead the SW-AETS January conference in Colorado Springs, with over 50 traveling in, some over icy roads to a day later basking in Colorado winter sunshine! Without delay, Paul then soon became a regional director on the AETS national board. Paul's professional involvement had been secured and his leadership at the national level had its beginnings.

As part of his initial work on the National Board of AETS, Paul was a participant in the evolution of the first journal for AETS, *The Journal of Science Teacher Education* (*JSTE*). Floyd Mattheis and others on the board had initially started the journal with David Haury as first editor.

The board of AETS at this time was full of like-minded folks who believed in the fledgling journal, bravely taken on by David, who was at that time a young professor at Ohio State in the Eisenhower Clearinghouse. In those days, *JSTE* was typically late in getting distributed because David was essentially putting it out from his home and on his own with an inexperienced editorial board. But it was published largely because of editor Haury's dedication, tenacity, and determination to do so.

To David's credit, he survived board criticism, and the board eventually learned that *JSTE* needed more than a dedicated single person to produce a quality journal. During Paul's tenure on the Board, he had the challenge of soliciting the call for a new editorial team. Fortunately for Paul and AETS, a group at Kansas State University led by John Staver came forward and Paul quickly recommended them for the 5-year challenge ahead. As that term was winding down, the board was approached by Kluwer in the Netherlands with a proposal to take over publication of the journal for a future editorial group. Paul traveled to the Netherlands and Belgium at his own expense to meet with the senior editor of Kluwer Publishing. He returned with information that made the board more comfortable signing a contract with Kluwer, thereby giving *JSTE* stature and recognition. It would fall to others on the board to select the next editorial team.

Simultaneously with the evolution of the journal, Paul, first as a SW-AETS director and then director at-large on the board, helped move the debate forward concerning AETS establishing its own annual meeting while also retaining connections to our then home group, NSTA. The debates about this move toward greater autonomy for AETS meetings were civil and rational with colleagues such as Kathleen O'Sullivan, Larry Enochs, George Davis, Norm Lederman, Jim Ellis, and the rest of the board. The debate and go-ahead ended with a wonderfully successful first independent AETS annual meeting in Charleston, South Carolina, in 1993. It was a significant risk, especially with hotel contracts and the threat of hellacious winter weather. But it proved to be a smart move by the association.

AETS was in a rapid growth mode in the 1990s. AETS President Peter Rubba and President-elect Norm Lederman recognized the need to develop a set of Standard Operating Procedures and then a set of Standards for the Development of Science Teacher Educators. During the early 1990s, Peter Rubba also led the organization in establishing updated bylaws. At the same time, Paul managed to convince colleagues that AETS needed a standard that spoke to our work with in-service teachers as well as preservice. Paul co-hosted with Peter 2 days of meetings in Denver, during which the first set of operating procedures were developed. Later, the standards would emerge somewhat in parallel with the NSES. Paul recalled Norm's look of surprise when he recommended that the document should include a standard that required future science teacher educators to have preparation in the field of professional development (PD). His own self-discovery of the research and literature in this area had convinced him that we should operate "professionally" in the area of in-service teacher development. Norm joined Paul's side of the argument, and the board adopted standards for the field of professional development for science teachers.

Serving first as a representative of the SW-AETS region then director at-large for AETS and all the while making a growing number of positive professional connections across the country, Paul was encouraged by Bobby Jones to run for president. After his initial declination, Paul put his cards on the table 2 years later and successfully ran for the highest AETS office in 1995. As with all then AETS presidents, this was a 3-year responsibility: president-elect, president, and past president. And so Paul Kuerbis' role as a leader in science education continued its upward trajectory.

Paul's years with AETS were solid and ongoing. He practiced multitasking well before the phrase became all too common; working on NCISE from 1989 to 1992, continuing to work with Jim Ellis on two successful NSF and Colorado Eisenhower proposals, continued appointment as a member of the National Science Education Standards working group on teaching, all while returning to his home department full time. There he would take on preparation of the departmental reaccreditation report for the state. He was especially proud of moving the department toward his vision of constructivist teacher education; the accreditation report and visit were successful.

With the opportunity of a full one-year sabbatical from CC from 1988 to 1989, it could be argued that it was at this time that Paul Kuerbis began his climb toward the pinnacle of his involvement and leadership at the highest national levels. So much was happening in Paul's professional life during the start of the 1990s that it is a challenge to pin it all down. Thus, we provide a Gantt Chart of key events in Paul's career that summarizes the highlights and overlaps (Table 11.1). However, one thing is certain: the results of his efforts, along with working with so many other talented people, are still in effect to this day. Most notably, this includes the production, and later evolution, of the *National Science Education Standards* (NRC, 1996), which were superseded by the *Next Generation Science Standards* (*NGSS*) published in 2012. A careful read of NSES, especially its emphasis on science conceptual ideas, inquiry instruction, assessment, and professional development can be seen in their early stages in the NCISE reports and even more visibly in the *National Science Education Standards*. Paul Kuerbis played an important part in the creation of those documents.

Members of the *National Science Education Standards* three working groups were all blindly nominated by at least two organizations. AETS was asked to nominate two and they selected Paul as one of their nominees, because of his work on NCISE and as a member of the board. Subsequently, the draft of the standards was created between 1992 and 1994. This first solid draft, some 40,000 copies, subsequently went out to many diverse groups for comment in 1994. When these comments were received, the National Research Council contracted with Angelo Collins, who had been a member of one of the working groups, to integrate everything into the final document released in early 1996. Paul and Angelo had worked together along with Rodger, Susan, Karen, and Audrey as an informal steering group that successfully argued for the need to set forth first a vision of inquiry teaching and then inquiry-based professional development. That would be followed by recommendations on assessment then those on science content. Paul's friendship and colleagueship with Angelo would continue in 2000 when Angelo would later ask Paul to serve as one of her advisors on a new venture as the founding executive director of the Knowles Science Teaching Foundation.

Finding a Sense of Balance: Being All Things to All Peoples

> Part of my deep and passionate involvement in science education at the local as well as national levels probably had to do with my ego or need to be in-

TABLE 11.1 Paul Kuerbis: Time Line of Major Events

Major Events	TIMELINES																																														
	1966				1970					1975					1980					1985					1990					1995					2000					2005					2010		2012
Oakwood School																																															
	■	■	■	■																																											
Berkeley				■	■	■	■																																								
Colorado College								■	■	■	■	■	■	■	■	■	■	■	■	■	■	■	■	■	■	■	■	■	■	■	■	■	■	■	■	■	■	■	■	■	■	■	■	■	■	■	■
BSCS																				■	■	■	■	■	■	■	■	■	■	■	■	■	■	■	■	■	■	■	■	■	■	■	■	■	■	■	
CAST															■	■	■	■	■	■	■	■	■	■	■	■	■	■	■	■	■	■	■	■	■	■	■	■	■	■	■	■	■	■	■		
AETS/ASTE																				■	■	■	■	■	■	■	■	■	■	■	■	■	■	■	■	■	■	■	■	■	■	■	■	■	■		
AETS Presidency																															■	■	■														
NSF Grants												■	■							■	■	■	■	■		■	■	■	■	■	■	■	■		■	■	■	■	■	■	■	■	■				
Colorado Grants																										■	■						■	■								■	■	■	■			
NCISE																																															
NRC																									■	■	■	■																			
Crown Center																																		■	■	■	■	■	■	■	■	■	■	■	■	■	■
Colket Center																																				■	■	■	■	■	■	■	■	■	■		
Knowles Advisory Board																																			■	■	■	■	■	■	■	■	■				
Knowles Trustee																																											■	■	■	■	■

KEY EVENTS IN THE PROFESSIONAL LIFE OF PAUL KUERBIS

Oakwood School	1966–1969
Berkeley/Lawrence Hall of Science	1969–1973
Colorado College	1973–2012
CAST	1980–2010
AETS/ASTE	
Regional	1974 then 1985–2010
National Board	1989–1998
Presidency	1996–1998
Conference Chair	2003–2005
NSF Grants	
Secondary Science/Math	1977
ENLIST Micros	1985–1986
ENLIST PD Materials	1986–1988
ENLIST PD Implementation	1989–1993
Co-STEP	1992–1997
CC-ISTEP	1994–1998
STEP-uP	2000–2008
Colorado Grants	
Eisenhower Secondary Science	1991–1992
STICS—elementary	1998–1999
STEP3—middle level MSP	2006–2010
BSCS	
ENLIST Micros	1985–1993
Elementary PD	1988–1989
Advisory Boards	1986–2011
NCISE	1988–1992
NRC	
NSES	1993–1996
CUSE	1997–2000
NSES and Inquiry	2000–2003
STEM Teacher Education	1997–2002
CROWN FACULTY CENTER	1999–2012
COLKET STUDENT LEARNING CENTER	2001–2010
KSTF	
Advisory Board	2000–2008
Trustee	2008–

> volved and the fact that I worked well with people in groups and they noticed. I guess these two traits led others around me to nominate me and ask me to be involved with the various national committees, running for the office of AETS President, and the like. I enjoyed being noticed like that, so it fueled my fires to continue. I can't really put my finger on a specific trait or was it a mix of self-enjoyment and working for the greater good? What I had realized in those years of the 80s and 90s was how I succeeded when I took advantage of opportunities, especially those that involved partnerships with other talented leader-types. (P. Kuerbis, personal communication, August 7, 2014)

While Paul's leadership in AETS (now ASTE) in the 1990s was of critical importance; his work with NCISE and NSSE and other projects also emerged as a result of this leadership. Paul really seemed to be adept at multitasking. Most individuals were not aware of his contributions to the many publications as both organizations had the culture of naming as authors all committee members in alphabetical order. Unfortunately, as a "K"uerbis, he was buried in the list of co-authors. And, in one major NRC publication on STEM teacher preparation, he was noted as a special consultant although he and the actual study director did much of the writing for the committee. Life in the fast lane can indeed have some unfortunate professional drawbacks.

Recall that Paul was "commanded" by his dean to return to his department full time in 1992, just as NCISE was winding down and NSSE was ramping up. Meanwhile, Paul and Jim Ellis, along with Paul's chemist friend and colleague Keith Kester, continued to collaborate on new projects based in Colorado and specifically at CC. In 1991 then, Jim (with Paul's contributions) produced a successful NSF grant, Colorado Science Teacher Enhancement Project (CO-STEP), with six sites across the state, one at Colorado College under Paul's direction. It built on lessons of professional development learned from ENLIST Micros. It was the first of several "STEP" projects from 1990 through 2008 that led many insiders to jokingly warn friends "don't step in it, but over it."

Simultaneously, Jim and Paul collaborated on a state Eisenhower proposal aimed at secondary science teachers. They would model inquiry pedagogy and use Paul's version of Bybee's 5E model—Paul's IDEA2 model—as a planning tool for the program, which ran for 9 months. Chemist colleague Keith Kester joined them. The teachers were invited and engaged by a field trip of the watershed of Colorado Springs and southeast Colorado on two all-day Saturdays in February. It was also a time for them to observe the landscape and discover the watershed. Teachers continued the explore and elaborate (in IDEA2), unpacking their observations and questions and forming special-interest groups to study more deeply such water-related issues as agricultural and recreation along with the overarching theme: "What is water quality for Colorado?" Teachers sampled water locally for "quality" (while learning how to use a number of HACH kits) and spent some summer weeks assessing quality and then establishing a case for defining quality appropriate for Colorado water. They then designed units of study, reflecting application of their new knowledge of content and inquiry pedagogy, which they implemented in the fall (Elaboration or Application in the model). They had experienced inquiry firsthand as well as constructivist learning theory in action, so their units of study were, indeed, novel and challenging for them to implement. Simply put, they all wanted more of Paul, Jim, and Keith.

This success led Paul to convince a dozen other science faculty that CC should start a new Integrated Natural Sciences degree to complement two other Master of Arts in Teaching degrees for experienced teachers: Humanities and Southwest Studies. It took months to develop a sound narrative, many conversations with STEM departments, and submission to the Committee on Instruction, which upon first hearing rejected it. The chair believed his science colleagues should be engaged in research. Paul persisted and resubmited along with a support letter signed by many STEM faculty. Then, at a monthly faculty meeting, the 8-page proposal was embedded in the agenda. The dean reminded the faculty of the proposal several times, but no questions surfaced, and it was approved unanimously. After the meeting, several colleagues complimented Paul on the careful work he had done in building support for the program and also support among non-STEM faculty who believed in the long standing College commitment to offer programs in support of K–12 teachers. Without funding by any agency, but with the wonderfully inexpensive summer tuition for experienced teachers—a long-standing tradition at the College—eight local teachers took a chance with Paul and his science colleagues. It was 1993 and the MAT Integrated Natural Sciences program was underway. This program was characterized by a 6-week content and modeling of inquiry pedagogy, six hours a day each of three summers. The summers were complemented by six Saturday all-day seminars for years 1–2 to further their work in implementing new pedagogy. They also engaged in classroom-based (Action) research studies and completed a culminating paper in year three and an oral "defense" about their growth and development over the 3 years. In this unique approach to a thesis, they needed to include evidence from their classroom work, as well as appropriate literature reviews supporting their journeys. Compiling one's journey into a substantial paper is, of course, a *reflective* approach.

During the first summer of Jim and Paul's Co-STEP grant, a program officer from NSF made a site visit. Over lunch, Paul ran the idea for NSF support for the INS program by the visiting reviewer from NSF. She was not enthusiastic. The following summer, 10 days before a due date for a teacher development program, Paul called the NSF visitor to chat again about his ideas. She pointed him to two specific pages in the Proposal Guidelines that might fit with what he had in mind. He alerted the campus Grants Office personnel, who informed him they had time to help. Quickly, he called Keith, who was on leave in Boulder working on surface chemistry, to come down to brainstorm a proposal. It was now 7 days before the Friday deadline. He and Keith roughed out a plan for a proposal on the Mierow House chalkboard as well as a financial model and how they could stage the project over 4 years. In the span of 5 days, while simultaneously teaching the elementary science teaching course in the MAT elementary licensure program, and running its associated children's *Whiz Bang* Science program, he worked with the CC grants office to create the narrative and a sound budget. The package was logged in at the Fed Ex office with 15 minutes to spare late on a Friday afternoon. It made it to its destination!

That project cleverly included funds to allow CC science faculty to engage for several days each summer at the College's beautiful Baca retreat center to become familiar with constructivist science, to inquire into a question themselves like children (Note: Keith and Paul used ideas from the 1960s ESS Mystery Powders unit), and then outlined a future 6-week-long full-time summer institute for the now-approved MAT INS degree program. But the main impact was on a cadre of K–12 teachers, about 10–20 each of the four summers, who were supported financially to pursue the MAT INS degree and take leadership roles in their schools. It also included a group from rural Colorado who lived on campus for 6 weeks each summer. The outside evaluator applauded them for modeling inquiry, deepening teachers' content understanding, and having year-long seminars that encourage and hold teachers accountable for implementing their new ideas and mutually support each other between three intensive summer institutes. The NSF funds ensured the long-term life of the program.

> NSF wanted us to focus on middle-level teachers and they agreed to allow others into the program on a space-available basis. The result was a K–12 audience of teachers of science. They quickly let us know how they valued the diversity, as they were able to talk to each other and gain appreciation for teaching no matter the grade level. It was a very positive time getting these populations of teachers to share insights and concerns with each other about teaching science to their own student cohorts. I should note, that everything we did during class was from research-based pedagogy. It was incredibly challenging to keep 3 to 4 cohorts working simultaneously. The first cohort of about 20 was manageable, but eventually we ended up with overlapping cohorts and a total of 40 one summer. Far too many to manage effectively. (P. Kuerbis, personal communication, August 7, 2014)

Paul's ASTE Leadership Wanes

Meanwhile, Paul had been visiting Washington, DC, several times a year for work on the NSES. Folks like Jay Labov at the NRC got to know him well and saw how he worked well with various groups and interacted with Rodger Bybee, who was then on leave from BSCS and directing the NRC Science Education center. Paul, as recent president of AETS, was contacted about serving a 3-year term as a teacher educator on one of two standing committees with the NRC: Committee on Undergraduate Science Education (CUSE). Dorothy Gabel, from NARST, also joined CUSE as the first teacher educators. CUSE was hard at work on a monograph about college-level teaching given the emerging implications for higher-education teaching based on implementation of the National Standards. Paul and Dorothy contributed greatly to the document, and Paul took the lead in authoring two chapters on Active Teaching and Learning—right up his alley from his work on NCISE and NSES. Again, his name was buried in the credits as simply a member of CUSE. But it was very rewarding to Paul to see how the committee's top scholars in STEM (from Stanford, the University of Chicago, and Princeton) wanted to bring about changes in higher education teaching and learning. Just as *Science Teaching Reconsidered* (NRC, 1997) was released, Paul successfully endured a significant and near fatal health issue, one that required many months of healing. But, as was typical of Paul's nature, he was right back at it as soon as the doctors permitted.

As he concluded his CUSE 3-year term, Paul was nominated by the NRC to serve on a study committee, co-directed by Jay Labov and visitor Jay Hackett (Paul's earth science colleague from the University of Northern Colorado), which focused on elaborating the meaning of inquiry in the NSES. Larry Lowery was also appointed, so it became like "old home week" for Paul and Larry each time the committee met. In the end, Paul helped craft a scenario for chapter five on preparing teachers to teach using inquiry. In fact, "Steve's Story" (pp. 88–90) is drawn from an MAT INS graduate's real classroom experience. Steve was a popular teacher and 15-year veteran STEM teacher who had no reason to change his approach to teaching physics. But after a year or so in the 3-year-long INS program, he finally changed his thinking and began fully implementing physics through inquiry. In this story, Steve flips his physics course upside down to one in which students primarily learn to design experiments to answer questions and only seek Steve's approval for proceeding realistically and safely in the lab. *Inquiry and the National Science Education Standards: A Guide for Teaching and Learning* was then released in 2000 (NRC, 2000).

But Paul's work within the NRC was not finished. In late1999, Paul had been coaxed by the dean of CC to become the second faculty director of the Crown Faculty Center and take full-time leave of the Education Department for an undefined period of time. Both thought it might become a transition for Paul so the department could begin to search for a new faculty member. It made some sense as Paul's long-standing chair was the same age and the dean knew a gradual shift would benefit the department.

Having made the commitment, Paul had to regretfully decline an invitation from Jay Labov on behalf of the NRC to take leave of CC for 2 years to direct a forthcoming study committee. This one would be charged to review the literature and best practices on STEM teacher preparation. In the end, Jay agreed that Paul would function like a study director with Jay as the official director and Paul as an unpaid Special Consultant. In the role, and with full time for his new faculty center directorship (normally only half time), Paul would gather all of the articles for the committee, respond to special requests from the co-chairs, and operate like a director. His role eventually became one of steering the committee to significant literature, co-planning committee meetings with the co-chairs, and then working closely with Jay and the chairs, drafting versions of the report for the committee approve. *Educating Teachers of Science, Mathematics, and Technology: New Practices for the New Millennium* was released in 2001 (NRC, 2001).

Working within AETS, implementing numerous grants, and co-authoring publications within the NRC, Paul stood out as a powerful influence on national policy agendas and outcomes. To his credit, he also did not lose track of influencing local and regional STEM teacher development. He gave back to the local community as much as he gave to national agenda's in STEM. He was also justly proud in admitting that he did not accomplish all of this alone, and he relishes the times and the people he worked with, was influenced by, and whom he influenced as well.

The New Millennium 2000 to Postretirement 2012

When he began his new work directing the College's Crown faculty center, it came after the then-dean, knowing of Paul's work on STEM professional development, thought that Paul's wisdom should be translated into helping his faculty colleagues, especially new-hires. It represented a shift, but one that seemed worth pursuing, and Paul certainly had cultivated many friends and colleagues among the entire faculty of about 160. With the luxury of being full time in what had been a half-time faculty position, the new dean encouraged Paul to submit a new NSF proposal for working with five local districts and their elementary teachers of science in 85 schools. His position would also allow Paul to take the lead in developing a centralized student learning center in the coming years. But, as luck would have it, we need to backtrack a few years to 1996.

At this time, with Paul as incoming AETS president, Colorado received a \$10 million systemic grant to impact K–12 STEM. It was written specifically to engage the public universities with a high-need district in their region. For Colorado Springs, this meant the local branch of the University of Colorado Springs (UCCS) and the Harrison School District with a very diverse population (economically and ethnically). Paul served on the state Advisory Board for the grant, assisting a bit in the development of the proposal and then more intensively once it was funded for a 5-year stint. In retrospect, Paul would reminisce that life in the Springs as a teacher educator came full circle from 1974 to 1996. In summer 1974, in addition to required teaching in his department, his STEM counterpart at UCCS invited Paul to join him in the Harrison district as their K–6 teachers became trained on various kit-based programs. Another year later, Paul would take Jack Sherman's place, with special permission, to teach an advanced STEM course. Two students went on for doctorates and have credited Paul with influencing their careers: Nancy Kellogg from Boulder and eventually director of NSELA—the new science supervisors group within NSTA; and Charles "Kip" Ault, who went on to have a rewarding career at Lewis and Clark College. Now in 1996, Harrison reached out to Paul to help them with their elementary science reform effort while UCCS continued to work with their middle and high school teachers. They were supported in this effort by the local branch of Hewlett-Packard. He worked closely with Linda Mooney, former principal, on professional development activities, as the district adopted various kits from the leading hands-on programs of the day such as FOSS, STC, and Insights. They formed a great partnership over 2 years and Hewlett-Packard nationally and locally encouraged them to consider a proposal to NSF for a "local systemic" initiative like the one the Palo Alto HP group had just received. Why not?

The proposal, known as STEP-uP, *Science Teacher Enhancement Program unifying Pikes Peak* region, received decent but not great reviews at NSF. In the second iteration, another district wrangled for control of the proposal. Some revisions were made and reviews were slightly more positive. Then, Linda and Paul collaborated on a successful proposal for state Eisenhower funds to implement a one-year version of the professional development goals contained in the NSF proposal. This served as a model of sound PD "modeling the model of inquiry," and all 47 teachers in the project signed a letter supporting the approach. The letter was included as part of the NSF proposal—a nearly $6 million request for a five-year project for K–6 teachers in 85 schools within five area districts. This was the third iteration of the proposal and followed a significant hour-long meeting between Paul and an NSF division director. This director shared comments from earlier reviewers and she listened to the ideas Paul and Linda were trying to implement. The revised proposal was submitted almost miraculously before the deadline (the reader is encouraged to ask Paul for the full story)!

A week after the review panels met in late October 1999, and while Paul was at NSF reviewing proposals for DUE, Linda received a call saying, "NSF wanted to fund us fully at just under $6 million and with no significant changes." Hewlett-Packard provided all the space for the project and it then became the poster child for Agilent Technologies (a new division of HP).

NSF did a visit in year three, applauded the work, and eventually suggested that Paul and Linda spend down funds into 2008. About 90 teachers completed the MAT-INS degree. Paul, full time at the College, served as a consultant (and Co-PI) for STEP-uP and burned the proverbial midnight oil to make it work. Over 8 years, the project impacted more than 2,500 K–6 elementary teachers of science. In an extensive regression analysis study, they were able to demonstrate that children who had a STEP-uP teacher in each of grades 2–5, and whose teachers also demonstrated implementation of "standards-based" teaching (small groups, questioning, use of notebooks, etc.) had a significant impact on children's performance on Colorado's standard testing: just over 1.0 standard deviation better than children not exposed to STEP-uP teachers for reading, writing, and mathematics. And in science, the difference was 1.8 standard deviations. The study was published in 2012 in the *Journal of Science Education in Japan* (Kuerbis & Revak, 2014).

As director of Crown, Paul now had to get up to speed on college-level faculty development, and in 1999 attended his first meeting of the Professional and Organizational Development (POD) network in higher education. An important new colleague was Michael Reder from Connecticut College, with whom Paul had shared many laughs and great brainstorming sessions around faculty development. The following year, Paul, Michael, and others formed a Special Interest Group of small college faculty developers. Almost immediately they and colleagues were running 3-hour workshops at annual POD conferences to help others dealing with the challenges of small-college teaching and learning. Paul quietly, or not so, embedded his IDEA 2 model in planning the 3-hour workshops, the last of which were done at POD and also at AACU. Of course, Paul was really enjoying his new role—helping faculty colleagues, mentoring new faculty, occasionally visiting classes, helping faculty engage in discovery of the emerging literature on learning and the cognitive sciences, and beginning to be involved in, yes, more grant writing! Over a 5-year period, working with campus colleagues and ones from other small colleges, Paul ensured that CC would be part of four separate, major studies and implementation efforts in higher-education reform funded (over $1 million) by the Teagle Foundation. Most of this work engaged faculty in understanding the importance of assessment and how data from surveys and from more direct measures might be useful to faculty as they *reflect* on how they might change the departmental curricula and their teaching. The work resulted in many presentations and half-day workshops at groups like POD and AACU.

Also in 1999, Angelo Collins approached Paul at a national meeting and showed him a *Chronicle* ad for an Executive Director of the Knowles Science Teaching Foundation. Neither had heard of the group. Was it legitimate? Later, they ended up brainstorming what a new foundation might do to change the preparation of STEM teachers. In the end, they decided that a missing component of teacher development was support during the first several years of a novice teacher's employment. The literature recognized the issue but no one seemed to be doing anything systematically to address it. Paul and others close to Angelo signed on as advisors, and Paul helped Angelo with the first professional development for these newbies—as part of the four initial "fellows" attending the NSTA in Portland, Oregon, in 2001. Later he and her advisors assisted in admitting successive cohorts of Fellows. In 2008, he was then nominated by Angelo to join the evolving Board of Trustees. Although Angelo had moved on to other ventures, Paul continued to serve the Foundation, now focused on identifying and supporting "leading teachers." Angelo's insight was reflected in the success of the Fellows and the outreach that took place by the alums, known as Senior Fellows. The keystone Fellows program then cited a retention rate of about 90% among these talented high school STEM teachers who hold the promise of leading their profession.

With Paul then on indefinite leave from teaching and education, a visiting position in STEM had been created within his home department. In order to continue the excellent work in the tradition of science education, Paul was instrumental in hiring two science educators to carry on the science education program. First was Paul Numedahl, who stepped in and served as Assistant Professor from 2002 through 2005. Then, after a national search, Peter Veronesi was selected as Visiting Associate Professor of Science Educator and served the Education Department during the 2005–2006 academic year after Paul Numedahl's departure. Finally, Mike Taber became the permanent science education replacement for Paul Kuerbis in 2006. All during the transitions of these three new faculty members who supported science education at Colorado College from 2002 on, Paul was always close by to contribute to the changes that were occurring and certainly to aid in brainstorming through any issues that arose. Yet Paul also gave each new faculty member the freedom to create the position as they saw fit. And for 2 years starting in 2008, the dean called upon Paul to step in as chair until Mike could achieve tenure and take the lead. This came as the College endured significant budget cuts, yet Paul persisted and prevailed in ensuring that the Education Department survived. Under Mike's capable leadership, it continued to thrive and a new integrated Education Studies major was on the books, serving many students interested in all things Education, but not necessarily in teacher licensure. Paul's former chair and someone who by chairing so long allowed Paul to thrive on his many projects, had then retired.

His return to Mierow House, a place were many ideas were generated, reminded Paul of how an upstairs classroom served as the focal point of much of his successful proposal brainstorming and writing. Most of the work with Jim and then with others occurred there. That place was also where former

AETS secretary Bill Baird worked on ENLIST Micros during the summers. And the living room downstairs was the site of Paul's summer AETS Board meeting in 1997. It was also there that Sandy Abell agreed to edit a new AETS Yearbook on International Science Teacher Education. Through it all, Mierow House seemed to have its magic with her as well as Sandy noted in the Acknowledgments for the volume:

> My guess is that, if I would have known then what I know now about what it takes to edit a book such as this, I might not have sat in Paul Kuerbis' little house on the Colorado College campus several years ago, so excited to start this project. (Abell, 2000)

Even while Paul Kuerbis became such a leader at the national level and was so successful at winning grants, perhaps a function of working many 7-day weeks, he still found time to mentor an ACE Fellow who joined the College in 2005 for a year. He and Susan Elrod worked closely together as she developed her STEM teacher development skills. Paul introduced her to the then director and founder of Project Kaleidoscope (PKAL), and in time Susan took on the role of PKAL director and helped merge PKAL with the Association of American Colleges and Universities (AACU). Paul served on the PKAL board from 2010 to 2014 in the role of helping guide their evolving role in higher education and with AACU. The reader may want to guess where the materials are housed for all of Paul's institutes. Yes, once again in Mierow House, this time in the basement!

In the fall of 1999, as Paul was beginning to direct the Crown Center, the College president, concerned with stagnant success rates among students (78%), charged the College with solving the problem. Paul co-chaired the Student Success Committee. The short story is that Paul played key roles in spearheading a 4-year effort to centralize student academic support services. A grant from the Priddy Foundation (which he helped write behind the scenes) led to the $1.1 million Learning Commons in Tutt Library in 2004. The Learning Commons now houses centers for both faculty and students in a space that overlooks the campus quad. Other aspects of the grant were used to help staff a coordinated effort for student assistance. And with another gift to the College from the Colket family, Paul led the establishment of a Quantitative Reasoning Center. As before, all of these efforts would eventually lead to a need for a directorship of the Learning Commons, a task that Paul was charged with in the early part of 2004 in addition to the Crown Center directorship. And, while Paul cannot be sure of the impact of centralized student support services, student success had climbed to almost 85% during his leadership.

Around 2003, the ASTE Board persuaded Paul to host its annual meeting in Colorado Springs. Others in Colorado are ready to assist so he says,

"Yes!" It is a team effort. Naturally he arranged for four days of 70-degree weather, perfect for some geology work in the field ahead of the conference and on its last day (Sunday). While some of the nearly 500 attendees became stuck in Midwest airports heading back to the east coast, he had many who asked him when he was going to retire. It wasn't in his immediate plans. He also had just opened his "new baby," the Learning Commons, and that needed his close nurturing and attention. After all, there may have been a few curmudgeon CC colleagues who might trip "the toddler" as it was just beginning to walk.

Officially retiring in late August 2012, Paul Kuerbis continued his role as a science educator by agreeing to help Mike Taber with yet another state grant and to work with a final cohort of 21 K–6 teachers from a local school district as they completed their MAT INS degree.

EPILOGUE

In rounding out his final responsibilities, Paul did eventually make a choice to back away from the centerstage at Colorado College and move to a "different place" for the enjoyment of his retirement years. In doing so, Paul Kuerbis leaves a legacy of hard work, dedication, and a bar that is raised quite high for those of us following in his footsteps. His career is also a template for those who find themselves the only STEM faculty member at smaller institutions. His wish is that we continually reach out to colleagues on our respective campuses and take advantage of all the opportunities that are available. Paul asks that we learn how to recognize these opportunities and then be intentional in your steps forward.

Pioneers of Science Education are volumes whose goal it is to document and celebrate the incredible work of many leaders in science education over the past many decades. What this author learned from reading about other pioneers and spending so much quality time documenting Paul Kuerbis' story is that each pioneer has innate personal attributes, characteristics, and dispositions that both drive them on to become successful in their own right. Persistence, enduring curiosity, and being goal driven are perhaps the traits that kept Paul on his train of science education. Indeed, it seems these traits were the ones so critical to his consistent advancement from his humble beginnings in "portables" as a new biology teacher to teaching professional development workshops and being the author of millions of dollars in National Science Foundation grants.

During his life and tenure as a science teacher education pioneer, Paul Kuerbis has seen a need, sought the resources to address this need, and was highly successful in the progressive development of science education opportunities for succeeding generations of science educators. Once again, yet another example of why we can see so far because we are standing on the shoulders of giants.

REFERENCES

Abell, S. K. (Ed.). 2000. *Science teacher education: An international perspective.* Boston, MA: Kluwer Academic.

American Association for the Advancement of Science (AAAS). (1989). *Science for all americans.* Washington, DC: Author.

American Association for the Advancement of Science (AAAS). (1993). *Benchmarks for science literacy.* Washington, DC: Author.

Kuerbis, P., & Revak, M. (2013). STEM teacher development in the United States: Key research recommendations and results from a longitudinal study of K-5 science teacher development on student achievement. *Journal of Science Education in Japan, 37*(2), 75–88. Retrieved from http://www.researchgate.net/publication/263654076_

National Center for Improving Science Education (NCISE). (1991). *The high stakes of high school science.* Washington, DC: Author.

National Research Council (NRC). (1996). *The National Science Education Standards.* Washington, DC: National Academies Press.

National Research Council (NRC). (1997). *Science teaching reconsidered. A handbook.* Washington, DC: National Academies Press.

National Research Council (NRC). (2000). *Inquiry and the National Science Education Standards. A guide for teaching and learning.* Washington, DC: National Academies Press.

National Research Council (NRC). (2001). *Educating teachers of science, mathematics and technology: New practices for the new millennium.* Washington, DC: National Academies Press.

National Commission on Excellence in Education (NCEE). (1983). *A nation at risk: The imperative for educational reform.* Washington, DC: Author.

ABOUT THE EDITORS

Jon E. Pedersen, PhD is a professor of science education, Associate Dean for Research in the College of Education and Human Science, and Director of Science Education for the Center for Mathematics, Science and Computer Education at the University of Nebraska-Lincoln. He received his BS in Agriculture (Animal Science and Nutrition) (1982), MEd in Administration, Curriculum & Instruction (Science Education) (1988), and PhD in Administration, Curriculum & Instruction (Science Education) (1990) from the University of Nebraska–Lincoln. Pedersen began his teaching career as a secondary school science teacher and taught high school chemistry and physics. Pedersen is very active in several national organizations including: the National Association for Research in Science Teaching (NARST), Association for the Education of Teachers of Science (Past President) (ASTE), and the National Science Teachers Association (NSTA). He is the author of over one hundred publications, most of which focus on science teaching and/or the incorporation of social issues into the extant curriculum. He has also published nine books and two teacher manuals. During his tenure in higher education at the University of Arkansas, in Fayetteville, AR, East Carolina University in Greenville, NC, The University of Oklahoma and The University of Nebraska–Lincoln, Pedersen has been primary investigator and co-primary investigator of numerous grants and supported projects on science curricula development, science in-service education, middle level education and international education totaling well over nine million dollars. Over the years, he has worked in more than a dozen different countries around the world.

Going Back for Our Future II, pages 227–227

Kevin D. Finson, PhD is a professor of science education and co-director of the Bradley University Center for STEM Education. He received his degrees from Kansas State University, including a BS in Science Education (1975), MS in Curriculum and Instruction with a focus on science (1980), and PhD in Science Education (1985). He began his career teaching earth and physical sciences at the high school level, then moved into teaching earth, life, and physical sciences at middle school. His college teaching has included earth science and geology, physical science, energy curriculum, teaching pedagogy and methods, and instructional theory. Among the professional organizations of which he is a member, Finson is very active in the Association for Science Teacher Education (ASTE). Other organizations in which he is active include the National Science Teachers Association (NSTA) and the National Earth Science Teachers Association (NESTA). He has authored over seventy journal articles, one book, several book chapters, and co-edited several books. A major focus of his work has been on students' perceptions of scientists (such as Draw-A-Scientist-Test research), and more recently on models of conceptions of scientists in learners' schema. He has also been active in publishing and work related to making science more accessible to students having special needs. Extramurally funded projects for which he has been a co-director include those on science and special needs, and development of science potential in middle school students.

Barbara S. Spector, PhD, a fellow of the American Association for the Advancement of Science and fellow of the National Institute for Science Education, is Professor of Science Education and Director of the Informal Science Institutions Environmental Education Graduate Certificate program at the University of South Florida. She received her degrees from Syracuse University: PhD in Science Education, a Master's degree in Combined Sciences and a Bachelor's degree in Biology. She holds a New York State Administration and Supervision certificate and taught high school biology for more than a decade. She has been awarded 67 grants totaling eight million dollars through which she initiated and maintained partnerships with professors in a variety colleges and universities, with informal education agencies, and with business and industry leading to development of 31 courses. Her research focuses on how change occurs in education. She developed and directed three major centers at USF, the STS Center, Project 2061 Research and Development Center and the Center for Ocean Science Education-Florida (COSEE-FL). She has been elected to national positions on the Boards of Directors of the key science education associations, has been active in policy making nationally and at the state level, has served on various advisory boards including for NASA and NSF, and been a consultant to a variety of state Boards of Education across the United States. She has been honored for her pioneering work introducing the use of social issues in teaching science through the interaction of science, technology, and society, and her long term leadership in the reform movement in science education.

Printed and bound by CPI Group (UK) Ltd, Croydon, CR0 4YY

15/09/2026

14962338-0002